THE BODY BLOG

Explorations in
Science and Culture

By

Rachel K. Wentz, PhD

ISBN-13:978-1530709724
ISBN-10:1530709725

Cover Image: Wound Man, courtesy of the Wellcome Library, London

Note on The Body Blog:

This book is based on a weekly blog written over a two-year period between 2013 and 2015. Imbedded in the blog are the reference websites used while researching each post. Therefore, this book lacks a traditional reference list. Referenced sites may be found by visiting the blog at www.rachelwentzbooks.blogspot.com and following the embedded links.

I have included the dates of each post within this book, since they track the seasons, events, and personal milestones I experienced during the years I wrote the blog.

THE BODY BLOG

Explorations in

Science and Culture

Other books by Rachel K. Wentz

Chasing Bones:
   An Archaeologist's Pursuit of Skeletons

Life and Death at Windover:
   Excavations of a 7,000-Year-Old Pond Cemetery

Let Burn:
   The Making and Breaking of a Firefighter/Paramedic

A Bioarchaeological Assessment of Health from Florida's Archaic

The Mass of Men

Searching Sand and Surf:
   The Origins of Archaeology in Florida

Table of Contents

THE BODY BLOG

Explorations in

Science and Culture

INTRODUCING THE BODY BLOG
April 3

When I was a child, my class took a field trip to St. Augustine's Fountain of Youth. Immersed in the humid heat of Florida's Atlantic coast, we drifted beneath ancient oaks, their limbs draped in Spanish moss, as our guide, sweating heavily beneath the bulk of his conquistador costume, wove the tales of that historic site. That trip would set a course for my life for it marked the beginning of my obsession with the human skeleton. But it also impacted me in another way: it was the first time I truly contemplated death.

Not that I wasn't already mesmerized by bones or naïve to the concept of death. My first experience with both occurred following the death of our obnoxious parrot, Polly, who finally expired after years of screeching and bad behavior. We buried him with great ceremony beneath a dark umbrella of palm and for days, I contemplated the changes taking place within the shoebox housing his remains, wondering where his annoying little soul was residing, whose hand he was now painfully pecking.

But it was in St. Augustine where I first contemplated my own death. After an exhausting day of trudging through history, our sweaty little group was shuffled into the final exhibit, which back then was an exposed burial ground where Native Americans had been interred some two thousand years before. I stared out over that burial, which held several adults and a few small children, and was shocked to realize that at one time, those pale, dry bones had been animated in life; that the skeletons that lay before me once belonged to living, breathing individuals just like me and that, like them, one day I too would die.

That experience launched my obsession. Suddenly, I viewed my body not merely as a collection of tissues that enabled me to go about my daily business, but as a magical vehicle through which I experienced the world. My body, with all its inner workings, was life itself. And like all of life, it would one day come to an end. I was hooked.

Since then, my careers—as a former firefighter-paramedic and as a bioarchaeologist—have revolved around the human body. As a medic, I dealt with devastating injuries and chronic illness, mitigating both as I frantically worked on patients, piecing together the fragmented remains of high-speed collisions, sending jolts of electricity through hearts ambushed by sudden cardiac arrest, and plugging the holes of those peppered by gunfire amidst the violent streets of Orlando.

As a bioarchaeologist, I examine the bones of the long dead in an attempt to interpret their life experiences: fractures resulting in angulated limbs, infection that settled blanket-like over the skeleton, dental disease that must have sent its host howling, and the creeping grip of arthritis resulting in gnarled and debilitated joints. This book is an extension of my obsession and a means of indulging two of my favorite preoccupations—the inner workings of the human body and how each one of us is impacted by the culture in which we live.

MUSINGS ON AN AUTOPSY
April 4

Let me go back in time, back to when I was a budding paramedic student, before I even knew there was such a field as bioarchaeology. Let me tell you about one of my early experiences with death, one that altered my concept of the "body" and forever changed how I view life.

I was in my first semester of paramedic school, barely twenty-two years old. I had already completed my emergency medical technician certification—a semester of coursework followed by months of fieldwork—and during that time, I had been exposed to the dead. But as an EMT student, I was relegated to the periphery. Suddenly, paramedic school thrust me to the forefront of patient care, where I was forced to make treatment decisions that would correct, stabilize, or kill my patients.

The paramedic curriculum is designed to help students come to grips with this new level of responsibility. The curriculum also forces students to confront death, so that they may develop the emotional callouses necessary for the situations they will invariably face in the field.

The highlight of my first semester was a trip to the county morgue, where we would spend the day observing autopsies. In reality, the visit served two purposes: not only were we forced to confront death in all its cold, antiseptic reality, but the autopsies would provide lessons in anatomy and physiology we simply couldn't get from a textbook or in a lab. This experience would leave a lasting impression on me.

The autopsy room was an assault on my senses. Bright lights glared against steel countertops, the sting of disinfectant penetrated my sinuses, and the sight of sheet-draped bodies made my stomach flip in nervous anticipation. I followed the tech to the nearest table. He slowly removed the sheet, folding it upon itself as he worked his way down the body of a young female. Her jaw lay slack, her eyes closed and sunken. Blood and cerebrospinal fluid seeped from her left ear and her fingers lay curled at her sides. Her fingernails were painted a frosty pink. Her toenails matched.

The tech explained the scenario. A waitress at an all-night café, the girl had been driving home from work early that morning when the driver's door in her pickup gave way. She wasn't wearing a seatbelt and was subsequently dumped onto the pavement where she landed on her head, suffering a fatal skull fracture. The tech reached for the gleaming scalpel and the autopsy began.

He made the typical Y-incision, starting at each collarbone, meeting at the sternum, and then drawing a deep straight line through her abdomen, unzipping her with smooth efficiency. He then peeled back her flesh and proceeded to give me a tour of her anatomy.

It was fascinating. Her glistening organs were neatly tucked within her belly, her heart and lungs nestled beneath her ribs. What captivated me even more than their organization was their stillness. The heart no longer beat, the lungs didn't inflate, her bowels had ceased their churning, and her stomach would never again growl.

He systematically removed each organ. Some were weighed, some were sampled by taking thin sections of tissue, which were deposited into small containers filled with formaldehyde. He squeezed the contents of her stomach into a jar as my own stomach lurched in response. After a few embarrassing dry heaves, I was able to regain my composure and the autopsy continued.

Despite its clinical setting, the procedure was strangely intimate. Her heart fit neatly in the technician's palm as he gently placed it on the scale. He carefully lifted a pale ovary from her pelvis and, with a delicate finger, pointed to the small shiny bulge that indicated she had been ovulating at the time of death.

When he made the large incision across the top of her head and folded her scalp down over her face, tucking it neatly beneath her chin, the woman no longer appeared human. Her body, lacking the animation of life, was but a collection of cells, tissues, and organs. She seemed no longer a person. She was now meat on the slab.

I thought about her often in the following weeks. I still think about her, decades later. That day in the morgue altered my perspective of life, death, and everything in between. An autopsy tears a person down to their foundation. It reduces the individual to his or her once-working parts and what is left is the stillness and silence of a lifeless machine.

But she taught me a valuable lesson. Now, when I peer into an ancient grave or examine the bones of those who lived thousands of years in the past, I force myself beyond their death to what they were in life. I think about their dreams, their fears, and how different those dreams and fears must have been from those we cultivate today. And, although their lives speak of a different time, we share a common thread. We are all descended from those ancient Africans whose bodies, through the slow accumulation of traits, would eventually become us.

We will explore that topic next.

MY COUSIN, THE LIZARD
April 7

Last weekend, a lizard took up residence in my kitchen. On nice days, I throw open the doors and bask in the glorious Florida sunshine. He must have sauntered in, taken a liking to the place, and decided to stay a while.

All day, we went through the same routine. I would enter the kitchen on silent feet, scanning every surface until I located that small splash of green. There he would sit, spread-eagle on the counter, leering at me with his reptilian gaze. I would lunge for him and he'd hightail it back to the safety of the alcove between fridge and stove. By evening, he was so emboldened that I found him lounging on the kitchen table, taking in the last orange of sunset streaming through the French doors. I made one last attempt as he skittered away, then I clicked off the light in disgust and headed for bed.

I didn't see him for a few days, but I knew he was still around. I found his calling cards: tiny black turds with the telltale white salt deposits. He was mocking me.

It wasn't until Wednesday night, as I was stepping into the shower, that I looked down and, lo and behold, there he was, peering at me like a peeping Tom from the base of the sink. I slammed the bathroom door, blockaded the jamb, threw a towel over the air vent, and gave chase. After a few minutes of awkward naked scrambling, I had him. I clutched his plump little body firmly between finger and thumb and headed for the front door. It was at that moment that I took a good look at him and couldn't help but notice our similarities.

He and I, like all vertebrates, share the same body plan, which evolved some five hundred million years ago during the Ordovician Period. Prior to the evolution of the vertebrates, the sea of life (and by that I literally mean "the sea," since it would require a spine for our ancestors to venture onto dry land) was populated by invertebrates. The spineless had the run of the planet for over two billion years before the vertebrates arose.

The earliest vertebrates were the jawless fishes. You are probably asking yourself, "What the hell is a jawless fish and how does one eat without a jaw?" Well, these strange creatures lacked movable lower jaws and were forced to scrape or suck their prey. The evolution of jaws made for more efficient eating, which led to more efficient hunting; a definite edge when your underwater neighborhood includes arthropods the size of NBA All-Stars. Two other necessities for life on land included lungs and limbs. In order to venture out of the water, it helped if you could breathe the air and get from point A to point B. Thus, the vertebrate body plan.

As vertebrates, we have a front and back and a top and bottom, in contrast to jellyfish, those beautiful, translucent blobs. We have a backbone that encases and protects our spinal cord, with a head at one end containing a brain. We have four limbs: two fore (arms) and two aft (legs), and we exhibit bilateral symmetry—our halves are mirror images of each other (at least on the outside; our guts are another story). These are the basics of the vertebrate body plan. Over millions of years of evolution, our bodies have morphed into what they are today. All the strange and wonderful adaptations we see throughout the phylum Chordata are accents on a basic plan.

What I realized as I carried my little friend through the house was that our similar body plans alluded to a shared history. He is a cousin—many million years removed—but a cousin all the same. Our bodies speak of our common lineage. And with that thought, I released him onto the front porch, and he disappeared into the night.

LEFTOVERS
April 14

Three years ago, my appendix burst. The pain started in early evening; an intense throbbing around my umbilicus. I assumed it was payback for the gargantuan Caesar salad I had scarfed a few hours earlier, so I threw back an Alka-Seltzer and went to bed.

But the pain intensified and spread throughout my belly, eventually forcing me to kneel before the porcelain and surrender the salad. I spent the night writhing in bed.

By morning, the pain had migrated south, spreading along the base of my abdomen. I assumed it was a virus, slowly working its way through my GI tract. I walked around clutching my belly for two days before finally giving in and driving myself to the ER. (I blame my former profession for the procrastination. In the arena of firefighting, there is no worse stigma than being deemed a candy-ass.)

After a quick scan, they diagnosed the problem, whisked me off to surgery, removed the offending organ, and I was on the slow path to recovery. The doc said I was lucky to survive. By the time they cut me open, the appendix was a shredded mess, my belly a mass of infection. Complications would send me back to the hospital a week later, but I eventually recovered, and the scar is barely noticeable.

My case is hardly unique. According to the University of Maryland Medical Center, over seventy thousand appendectomies are performed each year in the U.S. And even with modern medicine, thousands still die, primarily from lack of proper diagnosis and prompt treatment. But why is this? Why do we tote around an organ that has such potential to kill? And what the hell does this small appendage do, anyway?

For decades, we have heard that the appendix is most likely a vestigial organ—an evolutionary leftover. A body part that once functioned and, for some reason or another, has been dragged along the evolutionary path, even though it no longer serves a purpose.

We are not the only species that possesses such relics. There are cave-dwelling fish in South America that still retain their eyes, despite being blind and spending their entire lives in the dark. Remnants of leg bones can be found buried deep in the hind musculature of pythons and boas. And even some whales retain vestigial legs; a throwback to when they once walked on land.

We humans are no different. In fact, the appendix is not the only vestigial organ we possess. Evidence of our evolutionary journey can be found on our skin, in our mouths, and even on our skeleton.

Let's start with the skin. When the hairs on your body stand on end—what we commonly call goosebumps—you are exhibiting a carryover reaction from when our bodies were once covered with dense hair. By making the hair follicles erect, animals appear larger and more threatening. Think of a hissing cat in a defensive posture. And although we now lack the thick fur, the reaction persists.

Next come the teeth. Our dental arcade is reminiscent of the larger jaws of our ancestors, but over the course of evolution, our mouths have gotten smaller. Compare our jaws to those of the robust australopithecines; theirs were virtual nutcrackers, ours, not so much. Our smaller jaws can no longer accommodate a full arcade, which explains why dentists make such cushy livings extracting wisdom teeth.

Our skeletons also retain evidence of our primate kin. What once were tails have been reduced to a few small bones that taper from the end of our spines. The coccyx is but a mere nubbin of its former self.

Even some of our responses are programmed from an earlier era. For instance, you are walking down a dark alley when suddenly, a man steps out of an alcove wielding a knife. Your body automatically responds: heart rate jumps, respirations increase, and the pupils dilate as a hearty dose of adrenalin is dumped into your bloodstream. In the worst-case scenario, you wet yourself.

All of this is part of the fight-or-flight response, which is initiated by the amygdala, that region of your brain responsible for kicking your body into gear when you are faced with a perceived threat. This reaction and all its associated reflexes harken back to a time when we confronted physical danger on a regular basis. That same australopithecine was lucky to make it to adulthood without becoming a tasty morsel for some hungry leopard.

But back to the appendix. It turns out that little blob still has a purpose, even in our modern bodies. We now know it plays an important role in embryological development, producing compounds that maintain homeostasis within the developing fetus. In adults, it assists in the maturation of lymphocytes (white blood cells important for fighting infection) and even helps direct these cells to areas of the body in need of protection.

So, perhaps it is not so vestigial after all. Who knows, maybe we will discover one day that our coccyx has some magical power. Stay tuned!

DISFIGURED
April 21

When I was a child, I was accident-prone. During the three years we lived on the Philippine Islands, I managed to break my right arm, knock out two front teeth, tear the flesh from my foot, and nearly lose an eye. But the most serious accident happened when I was simply sitting in a chair.

It was one of those folding camp chairs. I was holding on to the scissor-like frame when it collapsed, snipping off the tip of my middle finger. My parents rushed me to the hospital (again) where a kind physician bandaged my finger using a small, mirrored splint. I stared in wonder at my maimed hand as my weary parents drove the familiar route home from the ER.

From that point on, I was self-conscious about my finger. Being right-handed made it difficult to hide, no matter how hard I tried. I felt everyone was looking at it, especially other kids. I carried it around in shame. It wasn't until the fifth grade that an incident made me come to terms with my unsightly appendage.

Her name was Ruth, and she had the misfortune of being born with only two fingers on her left hand. They weren't even normal fingers; they were claw-shaped and gnarled. Ruth rarely spoke and moved through life with her hand in her pocket.

Our class was assembling on the playground for that most barbaric of childhood sports, dodgeball. As we fanned out to form the circle, I noticed the other children, who were quickly moving away from Ruth, cutting their eyes and sharing knowing smirks. No one wanted to take Ruth's hand. She stood staring at the ground; her hand buried deep within her pocket.

Why is it we fear things that are different? Why do disfigurements or deformities cause us such discomfort, when we know those who possess them are no different on the inside?

Physical deformities have always fascinated me. As someone who specializes in the skeleton, I am most intrigued by skeletal disfigurements. They come in many forms. The process of creating a normal skeleton is riddled with complexity, the potential for defect, extensive. Conjoined twins, malformed limbs, bones that are too brittle, those that are too soft—there are a million ways building a skeleton can go haywire. All it takes is a hiccup in the timing, a glitch in the biochemistry, and what you get is skeletal chaos.

The structures within the human body are formed from three distinct collections of cells known as the primary germ layers. The skeleton originates from the middle of these layers, the mesoderm. This layer produces bone, cartilage, muscle, and the circulatory system that supports them. The ectoderm, or outer layer, forms skin, hair, nails, and the nervous system, among other things. And the endoderm, or inner layer, forms many of the body's internal structures, such as the respiratory and digestive tracts and the bladder. As these layers segregate and develop, they can be thrown off track by genetic abnormalities, illness or injury to the mother, or simply unsuitable conditions within the womb.

I have often tried to imagine what it would be like, living with a physical handicap. What does the world look like from the perspective of an achondroplastic dwarf, whose stunted limbs reduce him to a lower plane? What about those with severe scoliosis, where the spine zigzags down the back and the rest of the body is crippled from compensating? And imagine suffering from MOP. Myositis Ossificans Progressiva is one of those rare hereditary diseases that express themselves in frightening ways. Muscles and ligaments slowly ossify, encasing the victim within a bony tomb. As the muscles surrounding the ribcage harden, those controlling the jaw become fixed, and the patient can no longer breathe, no longer eat. There is no cure. There is no escape. In a world where such skeletal disorders exist, it seems a miracle so many of us possess normal, functioning frames. We are the lucky ones.

Let us return to the playground. I stood for a moment, self-consciously staring at my own finger, and glancing across the field at Ruth. And then I took a deep breath, strode over, and extended my maimed hand. Ruth slowly withdrew her hand from her pocket and tentatively placed her disfigured hand in mine, and together, we became part of the circle. Our hands were a perfect fit. Our disfigurements seemed to complement each other. From that point on, I tried to keep the oddity of my hand in perspective.

So, whenever you feel embarrassed about an aspect of your body, keep in mind there is always someone out there facing a greater challenge. To paraphrase Einstein: "It's all relative."

ANATOMY OF A FIRE TRUCK
April 28

Fire trucks are so cool. The next time you hear that ear-piercing wail of a siren, the bone-rattling blast of an air horn, take a moment to appreciate their beauty.

Referred to as engines by those in the biz, these high-tech monsters not only deliver personnel and equipment to emergency scenes, but they can turn night into day with elaborate lighting systems, supply heavy extrication tools via powerful on-board generators and, most importantly, deliver vast amounts of water to ripping house fires. They are mechanical marvels.

As a thirteen-year veteran of the fire service, I came to know these awesome machines intimately. As a firefighter, I rode backwards in a cramped jumpseat, sweating my ass off in summer, freezing that same tail in winter. As an engineer, I wove through the streets of downtown Orlando, barely missing the transients and tourists who leaped from my path. And as a lieutenant, I led an aggressive crew of professionals as we answered the endless emergency calls throughout the city.

Engines are the heart and soul of the fire service, and I can't help likening them to the human circulatory system (indulge me for a moment). Through an extensive network of hoses, an engine delivers vast amounts of vital water to awaiting crews, enabling them to extinguish flames, protect exposures, and prevent the spread of fire (you see where I'm going with this). In comparison, the human circulatory system moves blood throughout the body via its own extensive network, transporting oxygen and nutrients, fighting off pathogens, and regulating the body's temperature.

An engine typically carries an impressive five hundred gallons of water in its tank, whereas an average adult totes around five liters of blood, depending on that person's size and weight.

An engine depends on a centrifugal pump to move water, which is tucked beneath the engine mount and controlled by a gauge located at the pump panel on the side of the truck. The engineer adjusts the truck's RPMs to regulate the pressure within the hoselines. Too much pressure, the firefighters at the end of the hose become airborne. Too little pressure, their line goes limp.

Likewise, the heart, that fist-sized mass of striated muscle, pumps on average seventy beats per minute, which is over four thousand beats per hour, over one hundred thousand beats per day. Many factors affect the pressure within the body's vessels. Too much pressure can bring on stroke, heart attack, and internal damage to the system. Too little pressure and vessels collapse, organs fail, and the individual perishes.

Engines rely on frequent maintenance to keep their mechanisms functioning properly. Transmission fluid, oil, and brake and hydraulic fluid must be maintained for the engine and its pump to work effectively.

Similarly, the human body must be maintained to sustain a healthy heart. Lack of exercise can reduce the efficiency of the pump and, like the rest of the body, the heart becomes sluggish and lazy. Coronary artery disease and alcoholism can also affect the heart muscle, causing a condition known as cardiomyopathy, rendering it inoperable and, without a transplant, the patient may die.

Valves throughout the engine prevent a backflow of water within the engine's pumping system, while also regulating the amount of water that flows to each hoseline. The heart also has valves, four of them to be exact, which keep the blood flowing in the proper direction. Leaky valves can cause overwork within the heart, which causes blood to back up into the lungs, a condition commonly known as congestive heart failure. Valves also reside within the veins, those vessels that carry deoxygenated blood back to the heart. When these valves fail, blood can back up and engorge the vessels, creating that common condition in the legs: varicose veins.

An engine at a working house fire must have a supply line before it can provide multiple lines of attack. The line is laid from a hydrant and connected via an intake to the engine, providing an unlimited supply of water. Without a proper supply, the engine will quickly exhaust its tank, the hoseline will wither, the pump will cavitate, and the firefighters on the attack end will get cooked. Not a good scenario.

Without a proper blood supply, there is equivalent chaos in a human. Say an unrestrained driver is involved in a head-on collision. As his body is hurled against the steering wheel, his aorta—that large vessel that arches off the heart to supply the body—is torn. As the blood escapes into the chest cavity, pressure within the vessels drops, circulation to vital organs diminishes, the patient loses consciousness, and death quickly ensues. As with the dry engine, this is not a good scenario.

So, you see, fire engines and the human circulatory system have much in common. Both are mechanical wonders that deliver vital fluids to their constituents, both depend on a pump and hoses to deliver these fluids and, when operating properly, both are beautiful to behold.

THE BEAUTY OF FEET
May 5

I have goofy feet. Their goofiness resides on several levels. For starters, they are rather wide, thus earning me the nickname "Ducky" by my high school boyfriend, Patrick. I also had problems with my gait when I was young, which relegated me to a clunky pair of orthopedic shoes. I wore them for several years and cried and cried when I was forced to pair them with my new party dress.

And finally, I had the misfortune of acquiring the "bunion gene" (if there is such a thing) from my grandmother. The condition started rearing its ugly head when I was a teenager and was only aggravated by the high heels I favored when I had my heart set on becoming a model. Good thing I switched to firefighting. The footwear is far more practical.

With my bunions advancing, I knew I had to take action. I didn't want to end up with my grandmother's gnarled feet, so a few years back I had corrective surgery and now my feet are quite cute, despite the scars.

Every once in a while, stop and take a good look at your feet. When you think about it, they are quite astounding. Each foot is composed of twenty-six funky little bones that are connected via thirty-three joints and propelled by nineteen muscles. Over one hundred ligaments bind everything together to form a biomechanical wonder of locomotion. In fact, the fifty-two bones that make up the feet account for about a quarter of the body's total bony assemblage, as do the hands, which contain an additional bone each.

Did you know that per square centimeter, the soles of your feet contain more sweat glands than any other part of your body? The same can be said for the foot's sensory organs, which explains why feet are the perfect target for tickling.

Women tend to have more frequent foot problems than men and we can blame such sadists as Jimmy Choo and Christian Louboutin, whose ridiculous skyscrapers women feel compelled to clomp around in, despite the unnatural demands they place on the feet. A two-and-a-half-inch heel can increase the load on the forefoot by seventy-five percent. Just think what those six-inch stilettos do to our poor soles, not to mention our tempers. It's a wonder there aren't more stiletto stabbings in New York.

I guess it could be worse. We could still buy in to the ancient Chinese practice of foot binding. This strange cultural tradition, which may date back over two thousand years, would be undertaken when a young girl's foot bones were still pliable. Her feet would be wrapped tightly, and the toes and arches broken, in order to stuff them into teeny-tiny shoes. Bound feet were a status symbol and increased a woman's chance of finding a rich mate. This practice of creating "three-inch golden lotuses" was banned in 1912, although many women continued the practice in secret. It's a good thing it faded from fashion, for there would be no way I could get my wide "quackers" into such diminutive footwear. I would have been married off to the first rice farmer who came a callin'.

I must admit that high heels do enhance a woman's legs. They add height and length and force the calves to flex in order to remain perched atop the heel. Not so for those comfortable yet oh-so-hideous Birkenstocks. But then again, if you've given in to Birkenstocks, I doubt catching a man is high on your list of priorities.

I give credit to my down-to-earth sisters who refuse to put on airs, or high heels. I myself prefer a midrange heel, one high enough to appear feminine, yet low enough that I could outrun Jack the

Ripper, should he make chase. Although that stiletto might just come in handy. . .

THE ART OF MAN
May 12

It must be hard to be a man, especially in America, where "maleness" is defined by toughness, stoicism, and the ability to mask emotions. Having worked among firefighters for over a decade, I've witnessed maleness on a grand scale. But firefighters are a curious mix of machismo and compassion, for nowhere else do you have ultra-masculine males tasked with caring for complete strangers. I have seen men shift from verbal combat at the station to tearing up over the death of a child alongside the interstate, all within a matter of minutes. Males are a complex breed.

So how does one become a male? What is this process that separates the sexes? Let us take a look.

Each cell within the human body contains chromosomes, which house our genetic code in the form of DNA (deoxyribonucleic acid). Most cells of the body contain forty-six chromosomes. The sex cells, however—the egg from the female and the sperm from the male—possess only twenty-three. During fertilization, the egg, which contains an X chromosome, merges with the sperm, which contains either an X or a Y. The resultant fertilized egg, or zygote, ends up with a chromosome from each parent. XX and the zygote becomes a female; XY and you get a male. The Y chromosome contains a gene that directs the cells to create testosterone, thereby guiding the development of male sex organs. The X chromosome dictates the production of estrogen and, hence, female organs. Sounds simple, but it is really quite complicated.

Like any complex process, things can sometimes go awry. Occasionally the chromosomes do not segregate properly prior to fertilization, resulting in an odd number of chromosomes. For example, Klinefelter's Syndrome results in males who receive an extra X chromosome (XXY). These males are frequently sterile, with smaller than average testicles and larger than normal breasts.

Certain genetic disorders reside solely on the X or Y chromosomes. The sex-linked condition of hemophilia, which is found on the X chromosome, is typically expressed in males. (Yes, I know it sounds counterintuitive, but let me explain.) Females inherit an X chromosome from each parent; therefore, unless both parents pass down an affected gene, the unaffected X will dominate, and hemophilia will not be expressed. Males, with their single X chromosome, are more vulnerable. If their X is affected, they are born with hemophilia, since there is no normal X to override the defective chromosome.

Reproduction, like a good partnership, takes effort from both sides, but it is the males who call the shots when it comes to the sex of the child. It's those enthusiastic little sperm, and the X or Y tucked within them, that determine whether the squealing newborn will be a boy or a girl. If the process were left to us females, all births would be of the feminine persuasion. Males, on the other hand, are like blind gunmen: they simply fire their weapons and hope for the best.

So, the next time the male in your life stumbles over his pride or masks his emotions, try some patience. Remember that his maleness was programmed prior to birth and he is struggling against the weight of a culture that discourages tenderness. In the words of one of my favorite songs:

"I'm just the boy inside the man, not exactly who you think I am..."
"Be Somebody" by Thousand Foot Krutch

ODE TO THE LEECH
May 19

Here is a newsflash: Leeches are making a comeback!

Yes, it is true. Our little bloodsucking friend, banished from modern medicine for almost a century, has made a resurgence within the medical community. That age-old practice of leeching has swung back en vogue and its contribution to patient care and disease prevention is on the rise. So, let us take a closer look at these fascinating little suckers (sorry, I couldn't resist).

At what point in our evolutionary history did the leeches first make their appearance? It turns out they evolved from harmless freshwater worms, as evidenced from their DNA, as well as certain physical traits they share. They, like their wormy relatives, sport a sucker at the base of their tails that enables them to trundle along, inchworm fashion, as they make their way across the landscape or traverse potential hosts.

There are between seven hundred and one thousand species of leeches on the planet. The one most commonly used in medicine is aptly named *Hirudo medicinalis*. Many leeches live in water environments, both marine and estuarine, where they latch onto fish, thereby securing dinner as well as transportation. Those that scoot about on land prefer moist environments and are masters at camouflage, blending in with their surroundings as they search out their next meal. Not all of them feed on blood. Some prefer insects, mollusks, and even their humble cousin, the earthworm. Others are happy simply munching on surrounding ground litter. The bloodsuckers give them all a bad name.

I know what you are wondering. . . how do leeches make more leeches? It turns out leech sex is quite exotic. Many species are hermaphroditic, meaning they possess both male and female reproductive organs. Their procreative inventory includes several pairs of testes, one pair of ovaries, and, for those that engage in sex, a single orifice through which all the action takes place. They cuddle up to each other, either face to face or face to tail (who would have guessed they know about the "69" position?), and some don't cuddle at all; they simply deposit their sperm on their partner, which then dissolves a small portion of skin, allowing the sperm to enter the mate's body. Once inside, the sperm migrate to the ovaries, where they perform their magic.

And what role can they possibly play in medicine? An important one, it turns out, for they possess special gifts that only leeches can bestow upon needy patients.

Their application in medicine goes back over four thousand years, but they enjoyed their heyday in the 1st century and were employed for a range of issues, from fevers to farting. Galen, one of the earliest physicians, promoted the use of leeches to restore the "humors" of the body.

Practitioners achieved this through bloodletting: a technique of "bleeding" a patient to remove illness or impurities from the body. The leeches would be applied and would go to town on their host. And as it turns out, they are quite efficient. A leech can consume its weight in blood in a mere fifteen minutes! Pretty impressive. If given the time and an accommodating host, leeches can consume up to eight times their own body weight and, when full, they simply detach from their host and roll off.

The use of leeches was such a booming industry in France that over forty million were imported during the late 1800s. In fact, recently a French company was the first to request and receive FDA approval to market leeches for medicinal use. Let's face it, the French are simply enamored of the leech.

Thanks to a dedicated cadre of leech scientists, we now know more about their physiology and their potential application in modern medicine. It turns out leech saliva has magical properties. It contains natural anesthetics that help dull pain at the site of attachment, which comes in handy, for leeches are now being utilized to drain blood from swollen appendages following trauma and surgery. Leech saliva also dilates blood vessels, thus increasing blood flow to the site, which promotes healing. And finally, leech spit contains hirudin, which inhibits blood from clotting. This anticlotting property is now advancing the use of leeches on patients receiving grafts and reattached appendages.

For example, say you are slicing your morning bagel and accidentally lop off a finger. Once you arrive at the hospital and are whisked to surgery, the doctor reattaches the digit and you are wheeled to the recovery room. When you wake up, you look down to find a couple of chubby leeches affixed to your wounded hand and watch with morbid fascination as their bodies engorge to near-bursting.

Don't panic! They are doing what they do best. By applying leeches to newly grafted body parts, the leech's saliva keeps blood flowing to the area, thus promoting the growth of new vessels and enhancing circulation.

So, let's give the leech a little respect. It just goes to show that we can find a friend and ally among even our most humble relatives.

LOSIN' IT
May 26

It has been a hellacious week. Monday, I found out my job will disappear in October. My immediate reaction? Shiiiiiit!!!!!!!! But once I gathered myself together, I started planning.

The loss of my job may mean the loss of my house. Like many homeowners across America, I am still underwater on my mortgage, which means selling is not an option. The inability to sell means the loss of my down payment. Fortunately, I still have a nest egg, part of which I lost when the economy went to hell, but a goodly chunk remains.

Because my week has been a mad scramble, I was unable to prepare a proper post, so I beg your forgiveness. But coming to terms with my current roster of losses made me reflect on the losses we experience as we age. Here is a brief list that came to mind.

The loss starts with your belly button. When you are born, the doctor cuts the umbilical cord, leaving you with a diminutive stump that falls off after a few weeks. I remember when my baby brother lost his unsightly little nub. The loss continues with your teeth. Those tiny deciduous teeth you sprout as a toddler are eventually shoved out of your head to make room for the permanent dentition. If not, you would be severely challenged. Our larger permanent teeth enable us to bite, tear, and chew enough food to sustain our adult bodies, which would be impossible for the deciduous dentition. Besides, grownups would look rather freakish sporting baby teeth.

The loss increases with age, especially among our skeletons. We lose bone density, predisposing us to fractures and we lose cartilage in our joints, ushering in arthritis. Elsewhere, we lose vision as our lenses change and cataracts creep in; we lose our teeth, if we are not careful about oral hygiene; and we lose our mental clarity, as the synapses in our brains slow down and misfire.

And don't even get me started on appearances. Our tissues lose their elasticity, our skin loses its firmness, our tooth enamel loses its brilliance, our hair loses its luster, and our muscles lose their strength.

Life is about loss. Health is about trying to minimize that loss and beating back the forces of aging, such as gravity, inertia, stiffness, and weathering.

To put it all in perspective, I have much to be thankful for. Although fifty is creeping onto the horizon, I can still perform the same workouts I did as a firefighter in my twenties. Yes, I have required some general maintenance through the years, but my body is hanging in there and I am sure it will see me through to a new job.

So, do what you can to minimize your loss and keep the loss you experience in perspective. Remember, if you have your health, you have everything.

I FEEL YOUR PAIN
June 2

Humans can be amazing creatures. After last week's post on my impending unemployment, my friends responded with an upwelling of support and compassion. What was also amazing was that so many of my friends read my blog. Who'd a thunk it? Their kind words and thoughts got me thinking about empathy—that awareness that enables us to share each other's feelings and experiences.

In humans, empathy shows up at around the age of two, just as toddlers become self-aware. This newfound awareness is tested through the prodding of orifices, the male fascination with the penis, and the wonder youngsters experience when they are able to produce a booger. Even small children realize bodies are cool.

Humans are not the only species that expresses empathy. All mammals are capable of it to some degree, although it is more prevalent among animals that exhibit greater self-awareness. Elephants, dolphins, and especially primates exhibit empathy to greater degrees. You would be hard-pressed to find an empathetic camel.

Empathy was an important aspect in our evolution as a species. Without empathy, the ties that bind societies together would quickly unravel. Competition would eradicate the weak, ill, and disabled, resulting in a cold and self-serving world. (Why does Wall Street come to mind?)

Empathy is believed to be the key to altruistic behavior. Altruism—the willingness to sacrifice oneself for another—is found among many animals. Among Vervet monkeys, one brave soul will sound the alarm when a predator attacks, and although the alarm calls attention to himself and may increase his chance of being snatched, it provides a warning to others so that they may scatter and stay safe.

Not all forms of altruism involve immediate self-sacrifice. Among Florida scrub jays, hatchlings often stick around to help with the next batch, instead of venturing off to find mates of their own. Although this diminishes their chance of reproducing, it improves the chances of family members, thus perpetuating their own line. This is commonly called "kin selection" and is rampant among social insects. Among certain species of bees, only one lucky individual gets to breed. All the others are relegated to worker status and spend their lives toiling to maintain the hive.

Altruism seems counterintuitive to natural selection. Natural selection is supposed to favor the individual, whose personal attributes are such that they leave behind more offspring. You would think self-sacrificing behavior would be weeded out, since it decreases the altruistic individual's chance of reproducing. But it is believed that altruism is favored through natural selection at the group level, thereby promoting group survival and propagating the species.

It is difficult to imagine a world devoid of altruism. Altruism compels us to help the needy, donate our blood, and fight for our country. Even small acts can be altruistic: giving up your seat on a subway, holding your umbrella over an elderly person, or recycling your plastic. These acts may cost little but are what bind us together as a community, a country, and a species. And frankly, I would not want to live in a world bereft of altruism.

So, my dear friends, thank you for your concern. The messages laced with your kind words remind me that all of us share in the frets and fears of daily life. Whether it is losing a job, facing a diagnosis, or simply wondering what the next day might bring, our ability to empathize binds us together and will see us through.

TALKING HEADS
June 9

The young man was working as part of a survey crew when he was hit by the speeding truck. The truck's driver, tired of sitting in traffic, tore through the median, striking the young man and running over his body. I was a new medic; the patient was one of my first. He was alert and talking despite missing a large chunk of his forehead. The pale grayish surface of his frontal lobe glinted in the late afternoon sun.

During transport, I stabilized him as best I could—oxygen, cardiac monitor, two large-bore IVs—before concentrating on the open wound. I flushed it with saline and wrapped his head in a bulky bandage as we rushed full throttle to the trauma center.

His prognosis was grim. The surgeon explained they would have to remove a significant part of his frontal lobe to rid the area of grit and debris. The patient would probably never be the same.

This call was the first of many head injuries I treated during my career as a medic. It launched my fascination with brain trauma because, unlike other injuries, those to the head can result in a completely different person and steal one's identity—their personality, warmth, humor—leaving behind a shell of an individual.

I have witnessed a wide range of head injuries, both as a paramedic treating live patients and as a bioarchaeologist examining ancient skeletons. Working as a medic, the head injuries I encountered came in the form of shootings, crashes, and blunt force trauma. Barely a week went by without some significant trauma call, for the transients of Orlando's west side never passed up an opportunity to exchange thorough thumpings.

As a bioarchaeologist, I have examined head injuries among people who lived thousands of years in the past, and although these ancients lacked the modern weaponry of today's gang violence, their skulls indicate they pummeled each other on a regular basis.

It doesn't take much to damage the brain. Those sturdy containers our brains reside in are actually rather fragile. We tend to think of the skull as a singular bony compartment. Not so. The skull is composed of eight cranial bones that are joined via fibrous articulations known as sutures. There are four primary sutures of the cranial vault: the coronal, sagittal, lambdoid, and squamous. At birth, the sutures are not quite joined, thus the "soft spot" atop a newborn's head. But over time, the sutures fuse and are eventually obliterated, if the individual lives long enough. Bioarchaeologists use this "degree of suture closure" to obtain a rough age estimate at the time of death.

Fourteen bones make up the face. These bones are also joined via sutures; therefore, your head and face are a conglomeration of many flat and oddly shaped bones, all joined together to give you a distinct outward appearance. But back to the head. . .

Although the skull provides a protective shell for our fragile brains, simple injuries can have devastating effects. A fall from a moderate height, a minor car crash, even a love tap with a baseball bat can do severe damage to those delicate tissues. And, unlike other tissues in the body, nerves are unable to mend. Once they are damaged, they are done.

Luckily, our noggins contain about eighty-six billion neurons and, because the brain is wired with redundant circuits, sometimes even a serious head injury can be mitigated by the brain's ability to rewire existing connections. This is why physical therapy is so vital to head-injured patients, so that they can teach their brains new ways of accomplishing old tasks like walking, talking, and using their hands.

My patient was one of the lucky ones. Although the surgeons excised a large part of his frontal lobe, he suffered only minor memory loss and was otherwise fully functional. I got to know him during his long stay in recovery and it was amazing to sit and play cards with an individual whose brain I had come to know intimately.

So, take care with your cranium. A simple blow can produce frightening results. My brother fell off his bike years ago and suffered a subarachnoid bleed. He was restless and angry following the injury and, to this day, can be extremely temperamental. Although some of that may be genetic. . .

CRITTERS AND CONTAGION
June 16

Let me tell you the story of my encounter with a wild animal. It was the middle of the night and I was standing in my parent's kitchen, a glass of water in hand, looking out over the pool when I noticed the telltale ripples of a drowning critter. I headed out back, expecting to find the usual desperate frog, but as I approached the water's edge, I realized the ripples were far too big.

I peered through the darkness at a creature the size of a small dog. It was dark in color and kicking up quite a wake as it desperately paddled the shallow end. I crept forward and could just make out a small, pointy snout protruding above the waterline and a pair of clawed forefeet chopping at the surface.

It wasn't a frog. It wasn't a dog. It was a large armadillo that had somehow fallen in, his escape hampered by the thick shell encasing his body.

I grabbed a net and gave chase. He quickly veered toward the deep end, dropping a trail of turds to show his appreciation. I managed to corner him, then scooped him from the pool, flipping him onto the deck where he proceeded to hiss like a viper before turning tail to run.

Armadillos are a strange sort of creature, their anatomy an amalgam of many different critters. Their pointy snouts are rimmed by tiny sharp teeth, perfect for crunching insects, and their vision is quite limited, which explains why so many of them end up as roadkill. They sport long claws from their toes and have mangy tufts of hair on the undersides of their bodies. Their ears, like their snouts, are porcine and their shells are made up of dermal plates that enable some species to roll into a protective ball when threatened. They are related to sloths and anteaters and, I must say, are sorely lacking in personality. As if their cranky disposition and freakish bodies were not enough, they also hold the distinction of transmitting that most famous of biblical diseases: leprosy.

Leprosy, also known as Hansen's disease, is caused by *Mycobacterium leprae* and results in muscle weakness, nerve damage, and those terrible sores that spring to mind at the mention of this dreaded disease. Apparently, our friend the armadillo has recently been passing it on to us humans, especially in Texas and Louisiana, where it is hunted and, for some strange reason, eaten (don't ask me why).

But leprosy is just one of many zoonoses—diseases that have jumped from animals to humans. HIV is another. HIV is believed to have originated among chimpanzees in the West African rainforest of Cameroon. It is theorized that chimps contracted the disease by eating infected monkeys, then subsequently passed the virus on to the humans who ate them.

During the colonial period in the early 1900s, natives were conscripted to build railroads to facilitate the booming rubber industry. The natives received vaccinations within the labor camps and were inoculated using glass syringes. Unfortunately, there were not enough syringes to go around, and sterilization was not an option since autoclaves are hard to come by in the jungle.

Natives who fled the camps by escaping into the jungle were forced to eat chimps to survive. Many of those natives were caught, brought back to camp, and then inoculated along with everyone else, thus the spread of the disease from chimps to humans to more humans.

Other infectious diseases, such as swine flu and SARS, were also passed from animals to humans, and whether it is from eating infected flesh or living in close proximity to animals, germs have developed keen strategies for survival. Their ability to mutate and jump species is a testament to their fortitude.

But we should not blame the armadillos. Apparently, leprosy originated in the Old World and was part and parcel of European colonization of the Americas. Since armadillos are a New World species, they must have contracted it from humans and are now simply repaying the favor. And, in their defense, I'm sure the number of armadillos squashed by our cars far exceeds the number of humans they've bumped off via leprosy.

IT'S NICE TO SHARE
June 23

Last week, a ten-year-old girl received a lung transplant from an adult donor. What amazed me about this story—aside from the fact that surgeons were able to successfully fit a child with adult lungs—was that her case barely made news. Thus, a procedure once relegated to wishful thinking has now become commonplace.

According to the Organ Procurement and Transplantation Network, almost seven thousand transplants were performed in the U.S. between the months of January and March alone. So, let us take a brief stroll down memory lane and recount the history of these amazing procedures.

One of the oldest transplant stories is that legendary tale of Adam and Eve. According to scripture, Adam was gracious enough to donate a rib to procure himself a partner yet was betrayed when Eve's propensity for fruit landed them both in hot water. Transplantation was off to a rocky start.

True transplants began with that largest of organs, the skin. Hindu physicians back in the 6th century experimented by grafting the skin of a patient's arm and using it to reconstruct his nose. Apparently, the procedure was in great demand, thanks to the extreme judicial punishment of lopping off the noses of convicted thieves.

A similar technique was later adopted by the Italians in the 16th century, who would create a flap of skin, leave it attached to the arm until it sprouted the necessary blood vessels, and then use it to create the new nose.

It took another four hundred years for doctors to attempt kidney transplants. Unfortunately, they acquired the organs from an array of barnyard donors, namely pigs and goats. Not surprisingly, their patients did not survive, although I bet the practice kept hospital cafeterias well stocked.

In 1912, the French physician Alexis Carrell developed suture techniques he later used in the transplantation of blood vessels and organs, practicing on dogs, where he successfully transplanted kidneys from one pooch to another. He also perfected methods of keeping tissues and organs alive outside the body, famously nursing one of his samples along for over thirty years.

It wasn't until 1954 that a successful kidney transplant was performed on a human. The donor and recipient were identical twin brothers, Ronald and Richard, so rejection was not an issue. But it would take another six years for a British immunologist to receive the Nobel Prize for his discoveries in acquired immune tolerance, which opened the door for the creation of drugs that prevented rejection following transplants.

The 1960s also saw the first liver, pancreas, and lung transplants and, in 1967, the South African surgeon, Christiaan Barnard, made medical history by performing the first successful heart transplant. Unfortunately, the recipient, Louis Washkansky, died of pneumonia eighteen days later, but it was still a fine effort.

In the 1980s, Congress passed the National Organ Transplant Act, which addressed ethical issues of transplants (who gets what, when) and establishing a national registry for those awaiting organs. Folks at the Organ Procurement and Transplantation Network tell us there are currently over 118,000 anxious people awaiting organs, the majority keeping their fingers crossed for a kidney (over ninety-six thousand). Over three thousand are awaiting hearts, sixteen hundred are awaiting lungs, and forty-five truly desperate individuals are awaiting both. (And we complain about sitting in traffic.)

Today, transplant surgery has surpassed even sci-fi expectations. Facial transplants are becoming more common, providing victims of horrific trauma and devastating malformations the means to more normal lives. But as transplants become more common, the need for donors has skyrocketed.

This has bred a whole new industry: the illicit organ trade. In April, five people in Kosovo were arrested for running a clinic that procured kidneys from impoverished victims who came from as far away as Turkey to sell their organs. Donors were promised fifteen thousand euro; the recipients, mainly from Israel, were paying up to one hundred thousand for the transplants. Desperation lives at both ends of the organ trade spectrum.

In America, most organs come from those who die and leave their organs behind, and I am proud to be an official, card-carrying donor. I figure if I am hit by a train or carried off by a twister, they can salvage whatever is left and harvest some much-needed body parts. Granted, my organs are exposed to frequent doses of gin, but aside from that, they are in pretty good shape.

BODY ART
June 30

Have you noticed lately that tattoos are everywhere? You can barely walk down the street without encountering some form of body art. What was once relegated to military personnel and scary bikers is now common among teens and seniors. It seems everyone is sporting paint these days.

I doubt the trend is restricted to America. I was recently in London and did not recall such a plethora of paint, but then again, it was winter and naked flesh was hard to come by. I am betting they are inking up just like us.

Tattoos have a long and varied history. One of the earliest examples was discovered on the five-thousand-year-old desiccated body of the Tyrolean Ice Man, who was found in the Italian Alps where he died and froze (or perhaps froze and then died). He sported a total of fifty-seven tattoos, which were probably made by rubbing charcoal into lacerations made on his skin. Some of his tattoos may have been linked to a form of acupuncture since they parallel the traditional Chinese meridians.

The Egyptians practiced the art, since tattoos are found on their mummified remains and show up on artwork dating to over four thousand years ago. The Romans apparently did not share the Egyptians' enthusiasm, instead using tattoos to brand their criminals and it was not until they went up against the fierce Britons, whose painted bodies scared the bejesus out of them, that they came to appreciate the intimidating nature of tattoos.

But it was the Pacific Islanders who really set the trend. In fact, the word "tattoo" originated from the Tahitian word "tatau," meaning "to tap a mark into the body." The facial tattoos of Maori warriors were so impressive that a burgeoning trade in their decorated heads cropped up among Europeans in the early 1800s. The heads were in such demand that traders would capture members of the tribe, tattoo their faces, kill them, decapitate them, and trade the heads for guns. Ain't Colonialism grand?

Tattoos are not always permanent. Henna tattoos, which are created using the pigment of the *Lawsonia* tree, have also been in use for thousands of years across Africa, the Arabian Peninsula, and parts of Asia. The decorative painting of hands and feet is commonly tied to marriage customs. In fact, in India the longevity of the tattoos is believed to be directly tied to how well the bride will be treated by her in-laws. If the paint fades quickly, she is in for a rough ride.

My father, a captain in the Navy, sported numerous tattoos. His were gray and faded, but I can still picture the rooster inscribed on his calf: a talisman against drowning. When he underwent bypass surgery, the surgeon tasked with harvesting the veins from his leg was careful to suture the rooster back together. The bisection made the bird that much more interesting.

But tattoos are just one of many ways humans throughout the ages have adorned their bodies. Native Americans chose a more temporary route. They would create colorful paints from various plants and decorate their bodies with brilliant geometric designs. And they didn't stop with the skin. They would also use parts of animals—feathers, teeth, claws—to hang from their bodies, as a way of symbolically associating themselves with attributes of the creature.

Body art is tied to self-expression and has deep roots in human culture. Tattoos, like other forms of adornment, are a means of sending social signals. They say something about our beliefs, our world view, and how we view ourselves. As for me, I inked up a few years ago, once I had reached the point where my stint as a bioarchaeologist equaled the years I spent as an Orlando firefighter.

I had a gifted artist, Cesar, create a symbol of my two careers. What does it look like, you ask? It's a seven-thousand-year-old skull sporting my leather fire helmet and it exemplifies two of my greatest achievements: the grueling years I spent as a firefighter and the sweat and toil it took to obtain my PhD. I wear it with pride.

HOT AND COLD
July 7

America experienced a tragedy this week. No, it wasn't another school shooting, although it seems we can't go a day without some crazed individual exorcising his demons by inflicting mass casualties. This tragedy occurred when nineteen firefighters lost their lives battling a wildfire in Arizona. The wind shifted and they were gone.

Firefighter deaths are nothing new. Each year, the U.S. Fire Administration produces grim statistics that track the nature of line-of-duty deaths. Most of these deaths are not caused by fire; burns and suffocation make up a small percentage of firefighter fatalities. The majority are cardiac related, which is not surprising, since responding to emergency calls is incredibly stressful. For thirteen years I was jolted awake every third night of my life to answer calls throughout the city of Orlando. And let me tell you, going from zero to sixty when you are fast asleep is no picnic. Eventually you become conditioned, but it still takes its toll.

Fighting a building fire is also incredibly stressful. But wildfires make building fires seem like child's play. For one thing, you are working outside in the heat, usually in the middle of summer when fires are most active. And compared to the vastness of a forest, firefighters are simply specks on the landscape. We can only imagine how hard the members of the Granite Mountain Hotshots were working when they were overcome.

As an Orlando firefighter, I fought few wildfires. Most of the city has been paved over, so wildfires are few and far between. But it only takes one to realize they truly suck. With building fires, you can apply aggressive tactics to bring them quickly under control. With wildfires, it's you against Mother Nature. And we all know how hard it is to control a female. Any female.

Fortunately, firefighters are usually on the treating end of fire injuries and deaths. As a paramedic, I treated some horrific burns. From the young man who accidently doused himself with gasoline then walked too close to the acetylene torch, to a suicidal HIV patient who cut his wrists and then lit himself off, the burned patients from my past flash before me whenever I hear about a death by fire. They are not pretty.

As homeotherms (nerd-speak for warm blooded creatures), we rely on heat to maintain our body's metabolism. But it's a fine line between too much and too little. The average core temperature is 98.6 degrees Fahrenheit (that's 37 for those who use the Celsius scale; aka, the rest of the world). Temperature typically fluctuates about two degrees throughout the day, and age plays a role, with the elderly typically running a degree cooler than their younger cohorts. But what happens when temperature runs amuck? As you shall see, either extreme can spell death.

Thermoregulation—the ability to maintain a constant body temperature—begins in the brain, which is ironic, since it is the first organ to suffer when our temperatures spike. The hypothalamus, considered the body's thermostat, is made up of a cluster of neurons that are highly sensitive to changes in temperature, which it gauges via receptors in the skin and mucous membranes. When the core temp plunges, it kicks the body into gear in an effort to produce heat. Thus, your muscles contract involuntarily (shivering), and your body responds behaviorally, by rubbing your hands, moving your feet, and seeking an external heat source. Responses can also be gradual, as when summer wanes to fall. The hypothalamus triggers the release of hormones, which over time increase metabolism, subsequently increasing the amount of heat produced by the body.

Should these mechanisms fail or be unable to compensate (as when you are plunged into freezing water), hypothermia sets in quickly and consciousness is lost within fifteen minutes. As the core temperature dips below 95 degrees F, the body's regulatory systems cannot keep up and the internal organs pay the price. The circulatory and nervous systems quickly fail, and the body eventually expires.

At the opposite extreme, as in fever, that same thermostat in your head triggers dilation of the vessels beneath the skin (thus the flushed face of a febrile child) and sweating intensifies. Under extreme conditions, say running in the summertime heat, cramps can set in, especially in the legs. This can progress to heat exhaustion, which is accompanied by pale, moist skin, nausea, and vomiting. If left unchecked (body temp above 106 degrees), heat stroke ensues, leading to seizures, coma, and eventual death.

Each moment, your body is working; working to keep you warm, working to keep you cool. It is the body's innate ability to self-regulate that allows humans to occupy some of the harshest regions on Earth, like the Chukchi of Siberia or the Basarwa of the Kalahari.

But those brave men who died Sunday are a reminder of the fragility of our bodies and how susceptible they are to the brutal force of fire. For all it took was a change in the wind and a few fleeting moments. And a week later, America still mourns.

In Remembrance

Andrew Ashcraft, 29
Robert Caldwell, 23
Travis Carter, 31
Dustin Deford, 24
Christopher MacKenzie, 30
Eric Marsh, 43
Grant McKee, 21
Sean Misner, 26
Scott Norris, 28
Wade Parker, 22
John Percin, 24
Anthony Rose, 23
Jesse Steed, 36
Joe Thurston, 32
Travis Turbyfill, 27
William Warneke, 25
Clayton Whitted, 28
Kevin Woyjeck, 21
Garret Zuppiger, 27

DEAD AND BURIED
July 14

Last weekend, I attended a funeral. I'm one of those strange individuals who actually enjoys them. At funerals I can indulge my fascination with death, have the opportunity to view a corpse (which I also find thrilling), and witness the careful orchestration of remembrance. To me, these ceremonies are one of the most interesting aspects of culture.

I attended my first funeral while in grade school, at a time when I was just beginning to grapple with the concept of death. The service made a profound impression on me. Chuck was a beautiful boy in my sixth-grade class and the focus of an intense crush. We had just moved beyond gawking at each other over lunch and had progressed to the point of light conversation when he decided to go hunting with a friend. His friend tripped while carrying a shotgun, the gun discharged, and Chuck was killed instantly. I can still picture him in his casket. His dark suit couldn't hide the bulky dressing used to fill the gaping wound in his chest. I remember his blond hair.

Funerals have deep roots in human culture. For thousands of years, humans have celebrated death in some form or another. Even our cousins, the Neanderthals, incorporated ceremony into the disposal of the dead; at least, according to researchers at the Spanish site of *Sima de las Palomas*, where they have excavated the remains of at least six individuals who appear to have been intentionally interred.

But interment doesn't mean an actual funeral took place. That is where mortuary analysis comes in. By examining the myriad and often creative ways humans dispose of the dead, we can infer meaning behind such practices. First, we look at the evidence.

How is the body prepared? How is it positioned in the ground? What is included in the grave (grave goods)? And who is buried next to whom? These are some of the questions that provide the scaffolding on which we recreate burial rituals of the past.

Most archaeologists contend that intentional burial implies some form of ceremony. This is a logical assumption, especially when there are associated grave goods, the body has been treated or manipulated in some manner, and there are obvious patterns to the interments.

Interment, however, doesn't necessarily require ceremony, ritual, or even a belief in an afterlife. At its most basic, it is an economical means of disposing of a smelly corpse.

But as our brains evolved and our cortexes expanded, our cognition developed beyond mere self-awareness. With abstract thought came the ability to imagine, ponder, and reflect. Our large brains enabled us to see beyond the here and now and ask questions. Who are we? Where did we come from? And what happens to us when we die?

These questions needed answers. With imagination came the ability to conjure explanations for what we experience in life and what we just might experience in death. I believe the concept of an afterlife began as a simple explanation for that most vexing of questions: where do we go when we die?

If you have ever watched someone die, you can't help but wonder what is happening to them as they move through the process. I'm not talking about the physiology of death: the cessation of pulse, circulation, and breath. I'm talking about the more ephemeral aspects of a person. Their personality, their joys, their sorrows; everything that made that person who they were in life. Where does that go? Most people refuse to accept that the attributes that make each of us unique are simply based on the wiring of our brains, and when those wires stop transmitting, the person ceases to exist. Which may explain the concept of the soul.

Many religions treat death as a journey—to another place, another realm, sometimes even to a new and better life (depending on how well you behaved in the last one). And once people started viewing death as a journey, the next logical step was to prepare the deceased for the trip. Thus, adorning the body, gathering belongings, providing food, and, in extreme cases, sacrificing individuals to accompany the dead to the other side.

Of course, if you're an atheist like me, dead is dead. End of story. (I hope I'm right. I would hate to come back as a dung beetle.)

So, I attend funerals not for their spiritual or religious meaning, but with the curiosity of an anthropologist. But I also attend so that I can grieve and say goodbye, which is what funerals are all about. Even us nonbelievers appreciate a good sendoff.

THE UNSEEN
July 21

Late this afternoon, I found another tick. I have been plucking the little bastards off me for the past twenty-four hours, ever since I came in from the field. I write this not from the cozy comfort of my home, but from just outside Tomoka State Park. Located on the east coast of Florida and the site of my friend, Jon's, archaeological field school, the park contains numerous sand and shell mounds among which Native Americans lived thousands of years ago. A professor at Eastern Kentucky, Jon is here to investigate the mounds' histories and to train the next generation of archaeologists. Their first lesson in digging: wear plenty of bug repellent.

It's summer in Florida, which means three things: heat, humidity, and a plethora of insects. The woods of Tomoka host the usual buggy assemblage: spiders, mosquitos, and ticks galore. And it seems no matter how much precaution I take, I return to the hotel crawling with hitchhikers.

It makes me wonder how the natives survived. How did they make it through endless summers while infested with these nasty creatures? They were obviously a hardier breed than us, what with our air conditioning and screened porches.

But at least I can see and feel the ticks, for it turns out our bodies play host to myriad unseen creatures. Let's talk bacteria.

There are about two hundred different species of bacteria living on or within our bodies at any given moment. They colonize our skin, our orifices, and the different tracts of our bodies; those systems connected to the outer world via said orifices (think digestive and respiratory). In fact, bacteria outnumber our cells ten to one! And I thought a few ticks were bad.

Here are just a few of the bacteria that reside within or upon various parts of your body: in your nose, Corynebacterium mix and mingle; in your throat, Neisseria; your colon houses the enterics; and in your vagina (if you have one), reside the lactobacilli—the same genus used in cheese and yogurt (bet you'll never look at a cup of Yoplait the same again).

This entire assemblage is known as the microbiota and it has become a hot topic among biologists, geneticists, and just about every nerdy "-ist" in science. It's the new frontier in biological research and there is a particular set of science cowboys leading the charge.

The Human Microbiome Project (HMP), which is hosted by the National Institutes of Health, has set out to catalog the vast microbial communities each of us houses and to analyze the roles these critters play in our daily life and health. The HMP is taking a multipronged approach. It intends to not only identify and characterize the biome; it also hopes to sequence their genes in order to develop research strategies for investigating each and every species, with the hope of better understanding the role the biome plays in maintaining health and preventing disease.

Most of our bacteria are necessary for survival. If they stick to their assigned location and maintain their proper proportions, things tend to function normally. It's when they go roaming or their numbers fluctuate dramatically that the problems ensue.

For example, *Escherichia coli* (E. coli) is a type of bacteria that lives in your intestines. However, some types of E. coli, when ingested via contaminated food or water, can cause severe and sometimes deadly diarrhea. This is particularly troublesome in developing countries lacking proper sanitation (aka, a large chunk of the Third World).

Streptococcus is a bacterium often found on the skin or in the throat and in most cases causes no problems. However, pregnant women who are vaginally or rectally colonized (yes, I just used the term "rectally colonized") by group B streptococcus can pass it on to their newborns during delivery. This puts the babes at risk for developing meningitis, pneumonia, and septicemia. Sadly, these bacteria are one of the most common infectious causes of neonatal death.

On a positive note, studies of the microbiota can lead to fame and fortune. Just ask Barry Marshall and Robin Warren. In 2005, they earned a Nobel Prize for their discovery that peptic ulcers, previously attributed to stress and lifestyle, are actually caused by *Helicobacter pylori*, which commonly colonize the stomach but wreak havoc within certain individuals. I hope the prize winners celebrated over a giant greasy pizza.

If there is one takeaway from today's blog, it's that our bodies are so much more than meets the eye. They are teeming with life all their own, a world we cannot see but on which we depend. And for the most part, these bacteria are our friends. But keep your fingers out of your orifices and your hands clean, for bacteria are fine when they stay put, but can raise hell if they go awanderin'.

HOW WE KILL
July 28

Welcome to the great state of Texas! You remember Texas—our friends out west grappling with an onslaught of leprosy from feasting on armadillos. Well, last month, Texas carried out its five hundredth execution. And that's just since 1976, when the death penalty was reinstated after a brief hiatus. Who knows how many they've strung up over the last hundred years for such crimes as cow tipping and death by cactus. Texas is rough.

But Texas is not alone. There have been eighteen executions in the U.S. just this year (although almost half of these were in the Lone Star State) and capital punishment is alive and well in thirty-six out of fifty states. And how did we acquire such a lethal legacy? We can thank our friends across the pond.

The death penalty came ashore with the pilgrims and the first documented execution was Captain George Kendall's. A member of the Jamestown Colony in Virginia, Kendall was convicted of spying for Spain and was subsequently put to death in 1608. By 1612, Virginia's governor was so enamored of the practice that he enacted the Divine, Moral and Martial Laws, which allowed him to bump off any pilgrim caught committing the grievous crime of killing a chicken or stealing a grape. (And yet their state slogan is "Virginia is for lovers." Curious.)

Now here we are, four hundred years later, and although execution is typically reserved for only the most heinous of crimes, it is alive and well (no pun intended) in most states. Let's take a look at some of the creative methods we have devised to kill off such criminals.

The most popular method is lethal injection. It was adopted in Oklahoma in 1977 and the first to be tied to the gurney was a Mr.

Brooks in 1982. Strapping down the criminal is just the first step. Two IVs are then started (one is a backup; just in case) and a dose of sodium thiopental is given, which puts the person to sleep. Once he is resting comfortably, the prisoner is given a cocktail of pavulon and pancuronium bromide, which causes paralysis. Since inhalation is impossible without the use of the intercostal muscles, breathing is no longer an option. And, to cap off the exercise, a bit of potassium chloride is administered, which stops the heart. Done. And how many people have been "put to sleep" as they say in the veterinary biz? Over eleven hundred since 1976.

The next method (in order of popularity) is electrocution. The electric chair was born out of a need for a more humane form of execution than the old "let 'em swing" method. The first chair was constructed way back in 1888 and a Mr. Kemmler broke it in two years later. Typically, the person is shaved and then strapped to the chair. A large metal electrode is then attached to the scalp over a saline-moistened sponge. This improves conductivity (although if the sponge is too wet, you risk short-circuiting the machine). A second electrode is attached to the leg and the person is then zapped with up to two thousand volts. After the first shock, the pulse is checked. If the heart is still beating, another shock is given. The nastiest aspects of electrocution (aside from death) are the side effects: electrocution brings out the worst in people—I'm talking feces, vomit, and urine. To make matters worse, the limbs can contract with such force as to cause dislocations and fractures. It's not a pretty scene, but I wonder how it compares to what the criminal doled out to his victim.

On to the next method. In 1924, a Mr. Jon was the first to be put to death via cyanide gas, and the age of the gas chamber was born. Nevada was the first to use it, although their initial attempt was an embarrassing failure. They tried pumping the gas into Jon's cell while he slept, but the gas escaped, and Jon dreamt on. So, they hastily constructed a special chamber, which has been in use ever since. Once again, the person is strapped to a chair (I see a pattern here). Beneath the chair rests a bucket of sulfuric acid. At the signal, a lever is pulled, which drops crystals of sodium cyanide into the bucket and, through the miracle of chemistry, hydrogen cyanide is created. The prisoner is instructed to breathe deeply (no lie!) so that death comes quickly, although most tend to hold their breath and fight the chair. There is a specialized stethoscope attached to the patient through which a doctor outside the chamber can listen to the heart. When the procedure is over, the room is ventilated and the corpse sprayed down with ammonia, which neutralizes any remaining cyanide. Before the body is removed, the orderlies are instructed to "ruffle the victim's hair" in order to release any trapped gas (they actually have a "how-to" manual).

The final two methods are from a bygone era: firing squad and hanging, although they are still on the books in several states. Delaware, New Hampshire, and Washington still kill by hanging (although lethal injection is the primary method) and you can choose the firing squad in Oklahoma (and in Utah), but only in cases where lethal injection and electrocution are deemed unconstitutional (which is ironic, since it was the Okies who first introduced lethal injection, anyway).

Perhaps we should return to that old French favorite, the guillotine. Named after Dr. Joseph Ignace Guillotin (the "e" was added later), the contraption was invented by the kind doctor out of a desire for a more humane means of capital punishment that could serve all of humanity. Prior to its invention, the poor were usually "quartered" (for you non-Frenchies, that's a fancy term for having your limbs torn from your body). Hanging and beheading were reserved for the wealthy.

The first to benefit from Dr. Guillotin's creativity was a Monsieur Pelletie. On April 25, 1792, his head was the first to tumble (thus, the saying, "Heads will roll"). Thousands would follow in what became festive public events that drew huge crowds. Sadly, this fine custom came to an end—but not until 1977.

As you can see, hundreds of years' worth of blood, sweat, and ghoulish invention have gone into devising ways to kill, and capital punishment continues to be a controversial subject in the U.S. and abroad. If forced to choose, I'd have to take the easy way out: death by chocolate for me!

SCARS AND SOUVENIRS
August 4

This morning, out of curiosity (and possibly boredom), I tallied my scars. I did a head-to-toe survey, the way I was taught in paramedic school, and came up with a grand total of fifteen scars I have accumulated while strolling (and stumbling) down this road of life.

I find scars fascinating. They tell stories; they speak of the past. They are small (and sometimes gruesome) reminders of events in our lives. Some are accidental (where I tore my finger open on a hydrant during the fire academy); some intentional (the faint line in my belly button from a tubal ligation); and some are simply consequential, like the chicken pox scar on my forehead. They mark moments in life that are indelibly etched onto our bodies. So, what are scars and how are they different from normal tissue? Let's start with a scenario:

Say you are strolling down the beach and you fail to notice the broken bottle some drunken cretin had tossed into the surf. The bottle has now washed ashore, landing right in your flight path, and the next thing you know, your foot is laid open and you are hobbled. Two things happen. First, if you are like me, you curse like a sailor and proclaim your hatred for the common litterbug. Second, your body responds to the insult by forming a clot around the wound and sending fibroblasts, which help rebuild tissue, to the site. As the fibroblasts break down the clot, they replace it with proteins, primarily collagen, that will eventually form the scar tissue.

Scars are made up of the same tissue as normal skin; it's just the arrangement of the collagen fibers that sets them apart. In normal skin, the collagen is arranged randomly. In scar tissue, however, the fibers are generally aligned in a single direction, thus scars look and feel different from the surrounding tissue. And here is an interesting tidbit: scars lack hair, sweat glands, and a normal blood supply. Think of them as middle-aged men, except for that part about the sweating.

External scars are fairly benign. Although they lack the flexibility of normal tissue, they typically just dot our landscape and remind us of bygone events. It's when scar tissue forms on the inside of the body that the real trouble begins, and some of our most common diseases are linked to scarring.

Heart attacks (technically called myocardial infarctions) occur when a blockage in a heart vessel leads to death of the surrounding tissue. The damaged tissue can scar, which can then lead to arrhythmias (irregular beats or rates). Arrhythmias, such as atrial fibrillation, can subsequently lead to stroke. Scar tissue within the heart has also been linked to sudden cardiac death in people with certain preexisting conditions, such as cardiomyopathy.

Diabetes is also linked to scarring, as those who neglect to regulate the amount of sugar in their blood can suffer on multiple fronts. High sugar levels cause the smallest of blood vessels to burst, which can lead to scarring; and scarring around organs, such as the kidneys, can lead to failure, which is never a good thing.

One of the scariest diseases involving scarring is called scleroderma. Scleroderma results from an overproduction of collagen, those same fibers that create scars. These fibers can accumulate on or within the body, wreaking havoc along the way. Affected fingertips can develop pitting and sores and, in extreme cases, this leads to gangrene, requiring amputation. In the digestive system, scarring can cause acid reflux, along with poo extremes—alternating bouts of constipation and diarrhea. Scleroderma can even affect our sex lives. Men who suffer often experience erectile dysfunction, whereas in women, it can cause constriction of the vagina.

But scars can also be a thing of beauty. In fact, in many cultures, including ours, people intentionally scar their bodies. Scarification is a common practice among aboriginal populations in Africa and New Zealand, where they cut or brand their skin in decorative designs. For men, the scars intensify their appearance, which comes in handy in battle. But scars are also used as a means of attracting a mate – for both males and females.

In the U.S., scarification started cropping up in the 1980s, but it is also slowly spreading across Europe and Australia. It's achieved by cutting or burning intricate designs into the skin, much like tattooing. And like tattoos, these scars are used as a means of self-expression and can be quite beautiful. Scarification artists even have their own conference, aptly entitled, "Scar Wars."

Scars are friend and foe. They can produce deadly results when they intrude on healthy tissue but are also a means for the body to heal itself. They are a reminder of our experiences—both good and bad. So, flaunt your scars and share your stories, for who wants a perfect body, anyway?

The Body Blog

GUTS AND GLORY
August 11

I have a confession to make: I'm obsessed with roadkill. I can't even drive to work in the morning without gazing, mesmerized, at the previous night's slaughter. Living on the Florida coast provides ample opportunity, for there resides a plethora of potential victims. Raccoons, possums, and armadillos make for regular roadside fodder and their mangled, bloated bodies litter our otherwise scenic highways.

I've made a game of it. My favorite pastime, especially on long trips, is to try to identify the critter. Heads and tails are the most useful diagnostics, since these creatures are of similar size and their appearance tends to blur once decomposition sets in.

This hobby started when I was just a child and is grounded in my fascination with internal organs. I learned early on that roadkill afforded a rare glimpse at innards and I became so enamored of splattered guts that I once chased a large toad into the roadway and then crouched behind a bush until the inevitable car came along and flattened him. I quickly set to work inspecting the colorful smear; that is until my sister ran inside and tattled. My parents were not amused...

As one fascinated with guts, I chose the perfect profession. As a paramedic, I responded to a glut of horrific accidents involving mangled victims. Car crashes, stabbings, and shootings are just a few of the ways to lay open a body, revealing the beautiful, glistening entrails. In cases where the abdomen is intact, medics are taught to evaluate the belly by dividing it into four quadrants. For example, appendicitis pain typically flares in the right lower quadrant, issues of the spleen, in the upper left. By knowing where each internal structure resides, you can narrow down the suspect

59

organ, based on where the patient points. So, let's take a quick tour of our internal anatomy.

The human body is made up of various systems, such as the respiratory, digestive, and reproductive. These systems are composed of organs, which work together to perform certain functions. All humans have the same systems, with the exception of our sexual organs (unless of course you're a hermaphrodite and blessed with both male and female parts).

Organs are either hollow—those you can insert a finger into—or solid, which you can slice like a nice block of tofu. They both have unique attributes when it comes to injuries. Solid organs, like the liver, can bleed profusely when torn or bruised, since they house extensive networks of vessels. Hollow organs, on the other hand, such as the intestines, tend to leak their contents when damaged, which can lead to massive infection. It's pretty much a lose-lose situation when it comes to damaging your guts.

The largest of our internal organs is the previously mentioned liver. It weighs on average about two and a half pounds, is nestled just below the diaphragm on the right side of the body and is tasked with getting rid of toxins. The amazing thing about the liver is its ability to regenerate; if you remove a chunk, it can grow back!

But the liver is just one of many organs housed within our abdominal cavity. Its neighbors include the stomach, pancreas, and spleen, along with the intestines (large and small varieties) and those cute little kidneys in the back. In fact, those cute little kidneys are composed of about 140 miles of tubes and more than a million filters, and are able to cleanse the body's entire blood supply about twelve times per hour! Quite impressive.

You may be asking yourself, "What the hell is that spleen all about?" Yes, the spleen is a lesser-known organ – kind of like that poor astronaut who jumped onto the moon right after Neil Armstrong.

But the spleen is an important part of our immune system, storing white blood cells and getting rid of worn-out red ones.

What you may not realize is the important role the skeleton plays in the protection and suspension of these organs. First, there is our pelvis, which cradles our organs like a basin (and you thought your pelvis was only involved in walking and sitting). Our ribcage houses and protects our heart and lungs, but also those organs just beneath them (liver, spleen, and pancreas). And many of our organs are suspended via the peritoneum, which keeps them from sloshing around willy-nilly in our bellies. All of this is accomplished by our awesome bony frames and is just another example of the beautiful and intricate organization of the body.

So, you may abhor my lust for squished critters, and I'm not requiring you take it up as a pastime, but the next time you're confronted by a mangled roadside victim, take a peek. You just might enjoy it.

SKIN DEEP
August 18

I want you to think about your skin. Not simply a patch of it or the parts you don't like. I want you to picture it in its entirety. Start with your head and try to envision the one hundred thousand hair follicles that adorn your gourd (hairless individuals please proceed to the next paragraph).

Think about the skin on your face, how time etches its history on our features. If you're like me, you get a sinking feeling whenever you happen by one of those god-awful magnification mirrors (I really think those things should be outlawed).

Now picture the skin of your torso. If you're a fish-belly from up north, the skin on your abdomen is probably pretty pasty. For those of us living in the Sunbelt, we tend to accumulate a rash of freckles, especially on our shoulders, as we bake our way through life.

Proceed to your limbs. The skin on our arms tends to be of a tougher nature since it is frequently exposed to the elements. We are all familiar with the "farmer's tan" —bronzed forearms and a matching neck (thus that lovely term, "redneck"), and for centuries, women have caked their faces with powder to reaffirm they in no way participate in manual labor.

And finally, picture the skin on the palms of your hands and the soles of your feet. Those thick, waxy surfaces can't support hair follicles; thus, they are the only places on our bodies lacking hair.

The point of this exercise is to show that, although our skin is one continuous organ—yes, it is an organ and the largest one, to boot! — it is unique. Unlike other organs in the body, it has a wide range of textures, sports varying degrees of hairiness, and comes in a beautiful assortment of colors.

We have already discussed the embryological development of the skin in April's "Disfigured." As a quick review, skin arises from the ectoderm, that outermost germ layer from which many of the external features develop, such as hair and nails. The skin is composed of three layers: the epidermis (outermost layer), the dermis (middle layer), and the subcutaneous (you guessed it—the inner layer). The subcutaneous, also known as the hypodermis, contains the nerves, blood vessels, and the roots of our hair follicles.

This layering becomes significant in the cases of burns. A burn is categorized according to its depth. First degree burns are superficial and involve only the epidermis; sunburn is a good example. Second degree burns are those that affect the dermal layer and typically produce rubbery, fluid-filled blisters. Third degree burns, also known as "full thickness burns," penetrate the entire dermis. And finally, the ghastly (but rare) fourth degree burns, which involve muscle and bone.

Which kind of burn hurts the most, you ask? Surprisingly, it's the second degree variety, since full thickness burns typically destroy the nerves. The destruction of nerves means no pain. It's the periphery of the third degree burns, where third fades to second, that the pain resides. And we are talking major pain.

When we were kids, my sister suffered severe scalds to her lower legs when she climbed up onto the kitchen counter and spilled a pot of boiling water. She developed a whopper of a blister on her foot and I remember the fascinating way it would dimple when I pushed on it with my dainty finger. If only she had been a rhino. A rhino's skin is up to five centimeters thick. I bet it's hard to scald a rhino.

It could have been worse; she could have been a frog. Like many amphibians, our little froggy friends possess the unique ability to absorb water through their skin, which they utilize in place of drinking. They can also breathe through their skin. But, sadly, it's these amazing gifts that make them so vulnerable to pollution.

And here is a curious fact: polar bears—the most Caucasian of bears—actually sport black skin under all that white fur! (Word of caution: do NOT Google "naked polar bears"!)

Our skin is an amazing organ. It not only holds everything in place (for the most part), but it also protects us from pathogens, regulates our body temperature, and allows us to experience the world around us.

Imagine a world with no sensation, a world devoid of feeling. Think how important human contact is: the warmth of a hug, the thrill of a touch. Yes, we could do with less pain and suffering, but the pain reminds us we are alive. Pain and pleasure ground us to our world, and much of that pain and pleasure we experience through that wondrous medium, the skin.

TIMING IS EVERYTHING
August 24

The other morning, I burnt my eggs. I put six jumbos on to boil, anticipating the delight of an egg salad sandwich, but forgot to set the timer before sitting down to write. Forty minutes later, my concentration was broken by a loud popping sound emanating from the kitchen. I rushed to the stove as, one by one, my eggs exploded. As I scraped the pot, I got to thinking about timing.

Timing was a major theme in my previous occupation. The job of a firefighter-paramedic revolves around timing: how quickly you arrive on scene, how fast you get water on the fire, and how rapidly a patient is transported to the hospital. At OFD, even getting to our trucks was timed. We had to be responding to the call within one minute, or else! If you were in the shower or "taking care of business," too bad. In emergency services, timing is critical.

Now that I'm an archaeologist, timing takes on different dimensions. I no longer deal in minutes, my research spans centuries, and millennia. Thus, it was a shock to my system transitioning from the heart-pounding, adrenaline-pumping pace of emergency medicine to the tedious, methodical approach of archaeology.

As a firefighter, I experienced the joy of destruction. It's critical to open up confined spaces to search out hidden flames, so at every fire, we would blissfully tear out walls and pull down ceilings. God, it was fun.

Archaeology, on the other hand, requires painstaking excavation, the meticulous exposure of a site's history. My archaeological field school was an excruciating trial in patience. I imagined driving my fist through those perfect walls, taking a sledge to those stubborn soils. It's probably best I stick to the lab.

But it turns out timing is critical to my work as a bioarchaeologist, for the analysis of human remains is based on timing. Here's how.

For every skeleton that comes out of the ground, a basic assessment is conducted. This is the information that provides the fundamentals of bioarchaeological research. It all begins with sex (as everything should).

The sex of an individual is determined based on the size and shape of the bones, as well as certain landmarks on the skeleton. This is where timing comes in, for only after puberty do our skeletons develop the characteristic traits that distinguish male from female. The skull and pelvis morph with sexual maturity, making them useful tools in identifying the sex. Unfortunately, we lack a definitive means of sexing children, unless we are lucky enough to obtain DNA.

Determining the age at death is also based on timing. For children, the eruption of teeth provides a useful means, since we know, on average, when each tooth erupts in the mouth. For adults, we use certain surfaces on the skeleton, primarily the pelvis, because these surfaces change as we age.

Height is determined by measuring the long bones of the body, primarily the femur. The length of the bone is put into a regression formula, which reveals the overall height of the body. But once again, timing plays a role. Growth depends on the quality of the diet, exposure to disease, and, of course, genetics. But even if a child's growth is challenged, science has shown that children can experience "catch-up growth" in their teens, should conditions improve.

So why do we care about the sex, age, and height of individuals long dead? Because this information provides the tools with which we reconstruct past lives.

For example, let's say we excavate a cemetery of one hundred well-preserved skeletons (every bioarchaeologist's dream). If we know the sex of the individuals, we can deduce aspects of their social structure. If most of the skeletons are male, perhaps they were engaged in intensive warfare. An abundance of females may indicate differential access to high quality foods.

The age of the individual at death is also revealing. If the majority in our cemetery are children, it speaks to the overall fitness of the group since high child mortality does not bode well for a healthy population.

Height also provides information about overall health, for you are not going to grow big and strong if you are undernourished and plagued by pathogens.

So, even though I no longer treat patients or fight fires, timing is still fundamental to my existence. As the skeleton grows and matures, it changes, and it's these changes that provide the building blocks of bioarchaeological research by allowing us to tease information from the fragile remains of those who lived thousands of years in the past.

ERUPTION IN YOUR MOUTH
August 30

Date: Thanksgiving Day, early 1970s.

Location: Subic Bay, Philippines.

Subject: A budding tomboy, bored while awaiting the turkey, heads out to her favorite tree.

Event: While doing her best impression of a ring-tailed lemur, she loses her grip and plummets to the ground.

Status: Two front teeth, missing in action.

This scenario played out when I was just a child. The teeth were never found. Luckily, my permanent incisors came in strong and healthy, but this would be the beginning of a series of dental dramas.

I was born with an exceedingly small jaw. As my permanent teeth erupted, it was clear I lacked room for a full arcade, so by the time I entered my teens, it was time to take action.

They started with my premolars, yanking four of them to provide a bit of breathing room. Then came the braces; I'm talking the clunky metal kind that were oh-so-attractive. Once the unsightly apparatus was in place, the assistant set to work carefully chipping away the excess concrete from around the bands. As the tiny flecks hit the back of my throat, they triggered my gag reflex and the next thing I knew I was hurling a partially digested egg sandwich into that tiny sink. (FYI: Those little sinks can only accommodate spit.)

During my two-and-a-half-year tenure in braces, it was decided things were still too crowded, so out came my wisdom teeth. I've spent decades trying to convince my dentist I deserve a discount, since I lack a quarter of my inventory.

The final insult occurred on the racquetball court. I was kicking ass against two fellow firefighters when I took a racquet to the pie-hole, shattering my two front teeth. Fortunately, I had a gifted dentist who was able to reconstruct them, and today, no one's the wiser.

Teeth are important. If you've never stopped to appreciate them, take a moment. Imagine gnawing on a juicy steak or biting into a delicious apple. As I said, teeth are important. I bet you're wondering how teeth are formed. It just so happens I have the answer.

Tooth development, known as odontogenesis, begins in the womb and is controlled by over three hundred genes. By the fourth week of embryological development, the upper and lower dental arches are formed, which are covered by specialized cells that will eventually become the tooth buds; ten on top, ten on the bottom. The buds develop a cap of enamel, forming the crown, and dentin comprises the interior.

The deciduous, or primary, teeth are hidden within the jaws when we are born. The first to erupt are the two bottom incisors, usually when we are around six months old. Two top incisors follow, about two months later. The first molars usually erupt around our first birthday and the rest of the dentition quickly follows. Thus, you sport a beautiful arcade of twenty tiny teeth by the time you are three.

Then the permanents start shoving their way out. The twenty baby teeth will eventually be replaced by thirty-two permanent teeth.

Lower first molars usually take the lead, erupting around the age of six. The last to appear are the third molars, or wisdom teeth, usually in our late teens.

Permanent teeth come in four varieties: incisors, canines, premolars, and molars (front to back, respectively), and their shape is dictated by function. Incisors allow us to cut. Think back to that delicious apple. . . The canines are for tearing (or puncturing, if you are a vampire). The premolars serve a dual purpose: tearing and crushing. And the molars, with their broad, bumpy surfaces, are for grinding. Run your tongue along the chewing surfaces of your teeth and you will note the dramatic changes in shape.

Teeth hold information about our biological history. Dental traits are passed along as people move and mate. For example, the fascinating shovel-shaped incisors seen among many Native Americans were one of the first clues to their Asian origins.

Our teeth are also part of our evolutionary history. Over time they, along with our faces, have gotten smaller. Smaller jaws mean more crowding, thus the need for so many of us to have our wisdom teeth yanked.

But aside from the crowding, our teeth suit us to a tee. They allow us to chew, which aids digestion; they assist in speech by allowing us to form sounds; and they are key to good nutrition, for it's hard to be healthy if your teeth fall out of your head.

So, now that you know how teeth erupt, stay tuned. Next week we will discuss some of the bizarre ways we alter and adorn our teeth as we explore the exotic custom of dental mutilation.

BITING BLING
September 6

In 2006, after six grueling years of grad school, I finally graduated with my PhD. Having made the transformation from firefighter to ultra-nerd, I was quite impressed with myself. It was short lived.

I needed to find a job. But first, I needed to pack. I was part of a research team heading to Ukraine to investigate a two-thousand-year-old Scythian burial mound. As team bioarchaeologist, I would analyze the skeletons pulled from that ancient mound of earth.

At the time, I had only a vague notion of Ukrainian geography. I knew it was somewhere in Eastern Europe; a region that looms dark and mysterious to most Americans. As for the Scythians—I didn't even know who the heck they were. As a North American archaeologist, I had limited knowledge of classical archaeology (and between you and me, those classicists are a buncha' weirdoes).

But after a quick trip to the library and a crash course in Ukrainian history (including the fundamentals of the language: "hello," "thank you," and "where's the bathroom?"), I was ready to go. Twelve hours on a plane, fourteen by train, and a two-hour bus ride through the vast countryside of southern Ukraine brought me to the small farming community of Alexandropol. As I stepped off the crowded bus, I was met by our hostess, the woman on whose farm we would be living. Zena was stout, with massive, calloused hands and a demeanor that said she could squash me with her boot. But her smile stopped me in my tracks, for Zena sported a stunning gold grill. All her front teeth had been capped in gold. Had I wandered into some Ukrainian hip-hop backwater??

71

What I didn't know was that gold teeth are common in Eastern Europe—a sign of status, as well as a means of capping or replacing bad teeth. What I did know and could have explained to Zena, had I possessed the proper language skills, was that for thousands of years, people have been decorating their teeth for a variety of reasons, but mostly out of sheer vanity.

So, let's take a quick hop around the globe and sample some of the ways folks have blinged-out their mouths.

Let's begin with Africa (as everything does). Dental alteration, sometimes referred to as dental mutilation (you'll see why), goes back over fifteen hundred years in Africa. Seen in many regions as a rite of passage, the practice included the removal of certain teeth (using a stick or spear, or by knocking them out with a rock) or the chipping, incising, or reshaping of teeth.

Chipping the teeth meant knocking off edges to create a desired pattern. Incising the teeth—creating crosshatching or linear patterns—was done using sharp stone blades, like obsidian, or metal tools, when available. Reshaping was achieved by filing the teeth into various shapes, and for anyone who has ever suffered under the sadistic hands of a dentist, you can only imagine the pain involved in these traditions.

Extraction among children was also common. In parts of Uganda, a baby's canines were pulled, since it was believed infantile fevers originated in these teeth. This caused a rash of problems for the baby, including infection of the gums and malformation of the permanent teeth.

On the Indonesian island of Java, many adults participated in dental mutilation. They would file their incisors and canines, apparently as part of a long-held tradition. Engravings on the Buddhist temple,

Borobudur, which date back almost fifteen hundred years, appear to depict a person undergoing the mutilation process. The Javanese would also stain their teeth, which is ironic considering the obsession we Westerners have for blinding-white choppers.

The Maya of Central America also chipped and filed their way to beauty, but they are better known for their elegant inlays. They would drill small holes on the surfaces of their teeth and fill the holes with precious stones, such as turquoise and gold. The artists who created such dental dazzle used a small bow drill to create the hole before carefully placing the stones. The result? Looks that could kill (that's an inside joke—the Maya are famous for human sacrifice).

Today, we see bling of astronomical proportions. From rappers to swimmers, dental bling comes in a variety of gaudy styles; and, it seems, the more ostentatious, the better. Grills (at least the expensive kinds) are custom built to fit the teeth and are made of gold, rubies, and even diamonds, depending on the look you are going for.

Just blew your paycheck on a new tattoo? No problem. There's an economical alternative just for you. A quick Google search and I was able to locate gold-plated "Hip Hop Teeth Grillz" for a bargain $14.95! Awesome!!

But why should our teeth be any different from other body parts we adorn? We paint and pierce our way to fabulousness, so it only makes sense for our teeth to get in on the action. Humans are mad scientists when it comes to self-expression. And adorning, and even mutilating, our bodies are some of the strange ways we achieve it.

So, whenever my memory is jogged by an earthy fragrance, I think back to the wonderful experience that was Ukraine. I remember the stunning fields of sunflowers, the gracious generosity of the people, and Zena's beautiful golden smile.

Twilight now, we reach the shore
and raise the oaring high
move like dark familiar ghosts
against a grainy sky

We draw the boat upon the beach
ignite its brittle bones
and breathe the ash of smoldered wood
our shattered link to home

We claim a small deserted rock
proud kings upon the shore
and rule a dense, secluded patch
of ocean's level floor

And here we build a craggy nest
a dream among the sea
an island in a salted flood
a white eternity.

A NATURAL HISTORY OF THE PENIS
September 13

Ah, the penis, that most essential male organ. No other aspect of the male anatomy demands such attention. Stop for a moment and consider the plethora of names men have conjured for this illustrious male member. It's kind of like the Eskimo and their hundred words for snow. Priorities, I guess.

As a firefighter, I was surrounded by penises. They were everywhere. At the station, on the trucks, fighting fires—I was completely at ease waking up in a dorm full of woodies. So, let's take a closer look at this enigmatic appendage.

Sex, of course, is determined before birth by the joining of the sperm and egg and the blending of their associated chromosomes. X plus Y and Poof!—you have a boy. The recipe for a penis is embedded on the Y chromosome, but it takes about eleven weeks of gestation before the genital tuber (from which the penis will sprout) emerges. Ironically, at this stage in development, the genitalia of both sexes look about the same.

But by thirteen weeks, the penis is a penis. This is a lonely stage in its development, for the testes are still tucked away in the abdomen. They'll descend and join the party by the seventh or eighth month. The descent of the testes leaves behind a weakness in the abdominal wall, thus laying the groundwork for future hernias. So, if you are diagnosed with an inguinal hernia, you have your balls to blame.

We are all familiar with what the penis does. First off, it's a conduit for urine (let's get the dull stuff out of the way). The bladder empties into the ureters, which then dump into the urethra before traveling the length of the penis to exit. Since the urethra runs near the prostate, many men suffer from incontinence when the prostate

76

is damaged or removed. But urine is not the only fluid carried via the urethra. Let's talk semen.

The male reproductive system, at its most basic, is composed of the penis and testicles, but includes all the organs, ducts, and vessels that allow it to do what it does. Let's start with the testicles and work our way forward.

The scrotum is a fleshy sac that holds the testes. It is composed of skin and muscles and forms two side-by-side pouches: one testis per pouch. It's the smooth muscles of the scrotum that allow it to magically rise and fall. Body too hot? Scrotum drops. Body too cold? Scrotum draws up. Temperature extremes wreak havoc on sperm production, so it's the scrotum's job to keep things comfy.

The testes (testicles) are responsible for sperm production. They also produce testosterone, which is why boys destined for the opera in 18th century Italy were castrated. That way, they could continue singing in those beautiful falsetto voices.

Sperm produced in the testes then move into the epididymis—a network of thin tubules, several feet in length, that are bundled on the top of each testicle. This is where the sperm mature before being transferred up into the abdomen to the seminal vesicles, a pair of lumpy glands located on the backside of the bladder. The vesicles produce some of the liquid portion of the semen, which contains proteins and mucus, and has an alkaline pH. This enables the sperm to survive that oh-so-acidic environment of the vagina. And here's an interesting tidbit: the liquid also contains fructose, which provides a snack for the sperm should there be lag time while awaiting an egg.

The ductus deferens carries the semen from the vesicles and joins the urethra at the ejaculatory duct (things are getting interesting). It's this amazing little duct that propels the sperm up and out during ejaculation. In fact, it's so effective that it can propel semen up to two feet away (please don't try this at home)!

Enabling all the magic is the penis itself. This mesmerizing cylinder of flesh contains large pockets of erectile tissue which, when aroused, fill with blood, making it stand at attention. The erectile tissue also increases the size of the penis (goody!) and assists with insertion. Once inside, movement and friction along the shaft provide just enough gumption for the ejaculatory duct to do its thing and voilà! Ejaculation!

Recent research has shown that boys, like girls, are reaching sexual maturity earlier: by approximately two and a half months per decade since the 1800s. Scientists attribute it to better health and nutrition. As boys hit puberty, though, their chance of death increases, mainly due to higher levels of risky behavior. When testosterone production peaks, apparently you guys go bonkers and your tendencies for "dangerous and reckless shows of strength, negligence, and a high propensity to violence" increase your chances for fatal accidents; what researchers call the "accident hump" (let the snickering commence). So, for you youngsters out there, should you have the urge to jump off a building, try masturbating instead.

Lastly, we must address the issue of size. And how do our males fare when compared to our closest primate relatives? Quite well, actually. Compared to chimps, humans are much more endowed. Even gorillas can't measure up, although I'd pay good money to see a side-by-side comparison. And while we are talking size, let me clarify a misconception. Bigger is NOT always better. The female body can only accommodate so much girth before pleasure morphs into pain, so quit your bragging.

I will close with a fascinating bit of information from evolutionary biologist, Jerry Coyne. His website provides a world map of penis sizes, compiled by Dr. Eduardo Gomez de Diego. Keep in mind, lengths are based on self-reporting (and we know how you boys exaggerate!) and the sizes are in centimeters (didn't want the Americans to fall out of their chairs). So, take a look and see how you fare. Without going into detail, I'll simply say for my readers in Thailand and India: don't sweat it. As for my readers in the Congo —give me a call!

FROM ORAL TO ANAL: A GASTROINTESTINAL JOURNEY
September 20

Recently, I presented a lecture at Jacksonville's Museum of Science and History. My friend, Paul, is the new Curator of Education and after the lecture, he was kind enough to usher me through their wonderful exhibits.

To my utter delight, there was an excellent exhibit featuring the human body. We entered via a giant mouth and moved through a corridor lined with exhibits—an oversized cross section of the skin; videos featuring bloody surgical procedures; and my all-time favorite, preserved organs in jars! It was a nerd's paradise.

It also got me thinking. Every morsel of food we ingest follows the same pathway: in one end, out the other. So, I thought it would be fun to trace food's journey through the body. Come with me as we take a magical mystery tour through the gastrointestinal system!

We tend to think of digestion as the processing of food once it hits the stomach. But lo and behold, it is far more complex. The stage is set before we even take a bite.

Cutting up our food initiates the process. For starters, it makes it easier to cram into our mouths, but it also makes foodstuff more manageable. And once we take a bite, the magic begins.

Three primary components tackle the initial phase: teeth, tongue, and saliva. The teeth tear, grind, and crush, thereby prepping the food for digestion and enabling us to swallow. The tongue helps usher things along, moving the food around to facilitate mastication (a fancy word for chewing). The saliva moistens the food and starts breaking down carbohydrates, so as you are chewing, you are actually digesting!

The food then begins its long journey through the body. It leaves the mouth and enters the pharynx—that area between the mouth and esophagus. This is where things can get tricky, for we not only feed through this passage, but it also assists another vital bodily function: breathing. To guard the trachea (windpipe), a little flap called the epiglottis slams shut as food passes. We have all experienced the hacking that accompanies fluid "going down the wrong pipe." That is when the epiglottis falls asleep on the job and allows a trickle into the trachea.

Because of the tenuous positioning of the trachea and esophagus, the pharynx is ground zero for choking. If a bite is too large, it can get trapped in the pharynx, blocking the trachea, and making it impossible to breathe. My partner and I once pulled a ham hock the size of a golf ball from a guy's throat as we fought to restart his heart. Unfortunately, the ham hock won.

Food passes from the pharynx downward by the contraction (peristalsis) of the esophagus—a muscular tube that ushers the food into the stomach. Serving as gatekeeper is the cardiac sphincter, which closes off the stomach once food has been deposited.

The stomach is where the heavy lifting of digestion begins. It is a flexible sac, normally, about the size of two fists, filled with digestive enzymes and hydrochloric acid. Indigestion, commonly called acid reflux, occurs when the acid escapes upward through the sphincter, causing that common burning pain in the chest. (Personally, I've never experienced the joys of indigestion. I was lucky enough to be born with a cast-iron gut.)

As the food leaves the stomach, it passes into the small intestine. This magical coil of innards is about twenty-two feet long and takes up most of the space in your abdomen. Gory side note: I once worked on a guy who took a shotgun blast to the belly. When I got to him, he was fully conscious and lying in the garage, clutching a mountain of glistening guts that had been expelled following impact. Ah, the beauty of eviscerations—but I digress.

The small intestine contains ridges and folds which facilitate absorption of nutrients. As the food moves through this lengthy tube, other organs get in on the action. The liver produces bile, which it dumps into the intestine to digest lipids. The gallbladder picks up any excess bile and stores it for future use. And the pancreas provides a final squirt of enzymes to complete the digestive process.

Now comes the nasty biz of poop assemblage. The food enters the large intestine where a zealous colony of bacteria eagerly breaks it down, extracting the last bit of nutrients and magically transforming it into a turd. As we all know, feces come in a range of sizes, shapes, colors, and consistencies, so when you think about it, each bowel movement contains a delightful element of surprise.

When I left EMS to pursue archaeology, I thought my days of dealing with poop were behind me. But alas, it turns out paleofeces (aka ancient poo) are veritable treasure troves for archaeologists! They hold evidence about past diet, environment, and, on rare occasions, even contain the eggs of parasites. All that from a teeny tiny turd!

Prepare yourself for a squeamish sidebar: let's talk diarrhea. Diarrhea occurs when there is too much water in the stool. Is that it, you ask?? Hardly! One of the joys of writing this blog are the fascinating information trails I stumble upon during research. It turns out, diarrhea comes in several varieties, but by far the most fascinating and image-provoking has to be "chewing gum diarrhea"! No, it's not what you think. Chewing gum diarrhea is the result of consuming sugar substitutes such as sorbitol, commonly found in sugarless gum, which are not readily absorbed by the body. A side effect of this sweet treat is the overproduction of water in the large intestine, leading (quite literally!) to Hershey squirts.

The final stage in the digestive process is the elimination of waste. Since you are probably still reeling from the mental image of chewing gum diarrhea, I'll spare you the gory details. Let's just say that exodus occurs via the anal canal and the end result is defecation.

I hope you've enjoyed this titillating trip through the body. As you sit down for your next meal, think about the journey each bite is about to embark upon and pause for a moment of silent reflection at the wonder of the gastrointestinal tract.

HURTS SO GOOD
September 27

Have you ever been in pain? Not the annoying pain of a paper cut, or when your hammer misses the nail. I'm talking teeth-gnashing, stomach-churning, cursing-like-a-sailor pain.

Males and females have their own versions of ultimate pain. For females, it's the agony of childbirth; for males, getting racked in the testicles. Since I swore off children and lack a scrotum, I'm a virgin in both realms. But that doesn't mean I haven't experienced pain.

I suffered a few serious mishaps as a child (see April's *Disfigured*), but fortunately the pain associated with those injuries has been blurred by time. Since then, I've experienced true pain on only two occasions.

The first was the twelve hours I spent writhing in bed until my appendix finally burst (which ironically brought me a bit of relief). The second was a week later when the swelling in my belly caused my small intestine to crimp. Holy Mother of God!! It made the appendicitis seem like a day at the beach.

But what exactly is pain and why are we equipped with such reflexes? Let's explore.

Pain is a physiological response to noxious stimuli that warns us of danger. It comes in two forms: nociceptive and neuropathic. Nociceptive is what we typically think of as pain: nerves activated by insult or injury that send signals to the brain. Think of burns, trauma, or inflammation. Neuropathic pain is caused by damage to parts of the nervous system and is frequently chronic, meaning it can last for years. I can think of nothing worse.

Well, maybe one thing: the inability to feel pain. Known as congenital insensitivity to pain (CIP), this genetic disorder leaves its victims unable to distinguish true pain. The disorder affects the peripheral nervous system, leaving the person with a disconnect between the central nervous system and the nerves that detect sensation. The result is a shortened lifespan due to an accumulation of injuries and medical problems that go undetected. How would you know you were having a heart attack if you couldn't feel the chest pain?

Pain comes in many forms and the type or characteristic of pain can aid in diagnosing the underlying problem. Is the pain acute (sudden) or chronic (ongoing)? Radiating or non-radiating? Is it sharp or dull? Pinpoint or diffuse? Stabbing, piercing, throbbing, searing? The answers help weed out differential diagnoses.

As a medic working Orlando's west side, pain was part of my toolkit. The homeless are crafty when it comes to devising reasons to go to the hospital, where a warm bed and hot meal await (can you blame them?). Delivering noxious stimuli to transients feigning unconsciousness was part of every shift. We were masters at the sternal rub—knuckling someone's sternum to make them flinch. Only the most skillful actors evaded our assessments.

But what about mixing pleasure and pain? (You were hoping I'd go there.) I've never ventured into the dark closet of sadomasochism, but apparently there are a whole slew of folks who simply love the leather. So, I read up on the subject (I'm up Shit Creek if the FBI ever reviews my Google searches). According to *Psychology Today*, S&M is all about "power and control" and is far more prevalent than we realize. It comes down to role playing. The sadist wields the pain; the masochist is happy to receive it. Hey, as long as it's between two consenting adults, have at it.

Although we shun pain, it's vital to our survival. Imagine navigating life without pain's subtle reminders. Pain is a red flag for bad decisions. Think you can run a marathon? Give it a try and let me know what your muscles have to say about it. Going skydiving? Better train. I jumped during grad school (adrenaline withdrawals following my departure from the fire department) and hard landings were a reminder to use proper form. Pain is the mother hen of safety. Just imagining pain can sometimes make us pause and reconsider.

I'll leave you with a curious tale about pain (or lack thereof). A patient of mine, although drunk and with his leg in a cast, decided to take a ride on the back of his friend's motorcycle. Unbeknownst to him, the fracture had damaged certain nerve pathways in his foot. After a few joyful miles, he happened to look down and realize his foot had slipped from the pedal. He had been dragging his damaged appendage and subsequently ground off two of his toes.

Like I said, pain is a good thing. So, the next time you experience a mishap, remember that pain is your body's way of communicating. I strongly recommend you listen.

BELOW THE EQUATOR
October 4

As I brushed the dark earth aside, the first glimpse of a skull appeared. Slowly, a delicate profile emerged from the grave. I lay beside her—my arm extended, brush in hand—as I worked to expose her torso. She had been buried over thirteen hundred years ago, in a cemetery tucked among the dunes along the North Sea, in the shadow of the medieval castle, Bamburgh. I was midway through my PhD and had travelled to northern England to hone my skills as an archaeologist. Her cemetery was my classroom.

The shape of her bones indicated she was female: the high arch of her forehead, the slender angle of her jaw, and especially her pelvis, which exhibited the telltale architecture of a body designed for childbearing. In bioarchaeology, where we read the past through the skeletons left behind, the pelvis is one of the most effective means of assessing the sex of an individual. But it's the fleshed pelvis that really marks the difference between males and females.

We have already covered the male anatomy (see *A Natural History of the Penis*), so today we are zoning in on the female form. The subject is long overdue, for the female anatomy, though frequently splashed across screen and page, is the more mysterious of the sexes. Perhaps this will clarify things.

Although men and women look very different, they actually possess some of the same private parts. In fact, during certain stages of development, our genitalia are indistinguishable. Like men, women have a pair of gonads (ours are called "ovaries"), although unlike men, ours don't swing between our legs. The ovaries are tucked in the abdomen and connect to the uterus via the fallopian tubes.

The uterus is a flexible, muscular pouch that can expand to accommodate a growing fetus. It's also the source of those miserable cramps women suffer each month, as it sheds its inner lining (aka, menstruation). The uterus tapers downward into the cervix, which is connected to the outer world via that magical passageway, the vagina.

During the reproductive years, an egg is released each month by one of the ovaries. The egg enters the fallopian tube, makes its way toward the uterus, and the countdown begins. The egg has a seventy-two-hour window during which it can be fertilized. Sperm can survive for several days once they've been deposited— something to keep in mind for all you daredevils practicing the "rhythm" method of birth control. Side joke: What do they call people who practice the rhythm method? Parents!!

But back to the female anatomy. Since most women understand the fundamentals of their inner anatomy and most men couldn't care less, let's concentrate on the outer parts, for this is where men can lose their way. So, consider this as not just an anatomy lesson, but a means of improving your (and especially *her*) sex life.

It may come as a shock to you boys, but women actually have *three* holes down yonder and it behooves you to know the order in which they reside. The urethra sits just in front of the vagina, which in turn is positioned in front of the anus (another orifice you guys seem obsessed with). But most important for sexual arousal is the female version of the penis: the clitoris. Pay Attention!

The clitoris stands at the front of the line, just before the urethra (memorize the acronym CUVA: clitoris, urethra, vagina, anus). Like the penis, the clitoris contains erectile tissue, and, like the penis, it becomes stimulated when stroked. Arousal of the clitoris is fundamental to a woman achieving orgasm, so instead of being mesmerized by the boobies (although they require attention, too) or making a beeline for the vagina, you boys should be homing in on the "joy button." Take your time, do it right, and I guarantee you'll receive your just rewards.

As a cultural side note, you will be amused to discover that for thousands of years, the vagina has evoked fear and intimidation, resulting in outlandish legends concerning its dangers. The myth of the vagina dentata—literally a "vagina with teeth"—dates back to the Greeks and is rooted in the belief that the female sex contains hidden perils. Succumb to our charms and risk castration—or worse! Obviously, the stuff of fiction, but it pays to be careful.

I hope this little lesson has clarified that murky zone of the female genitalia. Yes, it can be tricky, but a little knowledge goes a long way. So, I wish you a safe and fruitful journey, and should you have the opportunity to exercise your newfound knowledge, be meticulous. And be sure to glove up!

DEAD AND LOVIN' IT
October 11

Have you ever seen a dead body? By the time we are adults, most of us have attended a funeral or two. It's a select few who go through life without actually viewing a corpse.

I've always had a fascination with death. From the time I was very young, I wondered about that thin line separating the living from the dead, and how, in an instant, life can be extinguished, leaving behind the shell of an individual.

I saw a lot of death as a medic. From homicides to suicides, accidents to natural causes, death was pervasive. And now, as a bioarchaeologist, I'm pretty much dependent upon the dead, since I rely on their skeletons to provide information about the past. I guess death and I are forever in collusion.

So, it's only natural for me to be curious about customs surrounding the dead. I've already touched on the theme of death in July's *Dead and Buried*, where I ruminated over the concept of burial. However, the range of funerary treatments is as vast and varied as culture itself, for death is the one aspect of life that demands universal participation. It's gonna get us all.

When the body dies, decomposition sets in, unless, of course, you are lucky enough to croak on a glacier. In forensics, decomp is broken down into two general stages. Autolysis occurs when fluids that normally reside in the intestinal tract are released and start digesting the body. Putrefaction follows as bacteria within the body reproduce unhindered. As the bacteria work, they release gases, thus the characteristic bloating. This is also the stage where insects descend since they find the gases irresistible.

It's the messy process of decomposition that has compelled most cultures to devise elaborate means of slowing or halting the process. Embalming, mummifying, and torching are all methods of avoiding a stinking corpse.

But I'll save funerary customs for another week. Today, let's talk sex. Not just your average, everyday sex—I'm talkin' sex with the dead.

Known formally as necrophilia, sex with the dead is not a modern construct. Even the ancient Egyptians knew to guard the bodies of beautiful women. They would hold onto them for a few days until decomp set in, just to ensure the bodies weren't diddled during embalming.

Today, psychologists identify three types of necrophilia. First, there is the harmless practice of fantasizing about the dead. These individuals typically don't act on their desires; they're usually satiated by deadly daydreams. Second are those who have access to a corpse and simply go for it. Makes you wonder about the quizzical grins of morticians. . . Third are the hardcore necros. These are the ones who kill in order to have a corpse at their disposal.

Necrophilia is a form of paraphilia: a condition where a person's sexual arousal and gratification are tied to abnormal or extreme behaviors. Think of them as the skydivers of sexual dysfunction. Paraphilias also include pedophilia, voyeurism, and S&M, among others. As prevalent a role as sex plays in culture, it's no wonder we have a whole slew of deviancies associated with it.

On rare occasions, necrophilia is partnered with other dysfunctions. Take Jeffrey Dahmer, for example. Not only did Mr. Dahmer kill, mutilate, and violate his young male victims (typically in that order), he also served them up for dinner. Jeffrey was indicted on seventeen counts of murder after one of his potential victims made a hasty escape. The boy managed to flag down a few officers, whom he then led back to Dahmer's lair. There, they discovered human heads scattered throughout the apartment, a plethora of body parts in the fridge, and a photo montage depicting his gruesome hobby. His arrest brought an end to his grisly pastime, but you can imagine what the cleanup entailed.

Fun fact: the term for ingesting the flesh of the dead is necrophagia. Cannibals, on the other hand, tend to prefer fresh meat.

Psychoanalyst Erich Fromm mused on the character traits of necrophiliacs in his book, *The Anatomy of Human Destructiveness*. He viewed these individuals as products of social evolution and listed their personality traits as follows: Use of language that includes numerous death-related words, dreams involving death or dead parts, and an interest in sickness. Holy Sh#t!! He just described the entire audience at our last bioarchaeology conference!

To avoid being cast as a necrophiliac and to assure you I have no intention of "digging up" my next date, I'll close by saying that necrophilia encompasses a wide range of dysfunction, but not necessarily those of us in the skeletal biz. I've never been sexually aroused by the dead, despite being surrounded by an abundance of bone.

So, on behalf of all bioarchaeologists, I proudly proclaim that, although we are fixated on the dead, we are not out there lookin' to get lucky.

SOMETHING SMELLS
October 18

Close your eyes and breathe deeply through your nose. What do you smell? Depending on where you are while you are reading this, the range of odors may be vast. Sitting in your office? Perhaps you smell the ink from the copy machine or a pot of coffee brewing down the hall. At home, as I am at this moment? Perhaps you smell fresh laundry tumbling in the dryer or the wonderful attic smell of an old house. If you are sitting on the john, we'd rather not know what you're smelling at the moment.

Smell is one of our most important senses, although we rarely give it much thought. When I lived in London, my morning commute on the underground was a banquet of human odors. The diversity of people on the subway meant a broad swath of scents: spices, fragrances, and body odor. A veritable smorgasbord of aromas.

For firefighters, smell is part of the toolkit. Ask any firefighter and I bet they can describe the acrid odor of car fires, the earthy scent of a brushfire, or the stomach-churning stench of burnt flesh. The smell of the smoke at times can indicate precisely what is burning. So, today, in honor of our snouts, I'm paying homage to this vital link to the world around us.

First, let's discuss the nuts and bolts of smell. Our noses serve as conduits for odor molecules that are drawn into our bodies as we inhale. Those odor molecules attach to chemoreceptors that line our noses, which, when stimulated, send signals to our brains. The brain then interprets these signals and voilà! A smell!

Humans can distinguish over ten thousand different odors. Amazing! Also amazing is that each of the hundreds of receptors lining the nose is controlled by a single gene! If you are missing that gene, you miss out on that particular odor.

The olfactory bulb—that part of the brain that transforms smell sensation into perception—is part of the limbic system, a primitive region of the brain that regulates behaviors related to survival. Even in our distant reptilian relatives, the alligators, the limbic system processes smell, which allows gators to successfully hunt and defend their territory.

But it's the amygdala and hippocampus, the parts of our limbic system responsible for emotion and memory, that draw the connection between smell and memory. The olfactory nerves run in close proximity to these vital areas, thus providing associated links between smells and the memories they trigger.

As our species evolved, our sense of smell would have played a crucial role in survival. Finding food and avoiding danger are two of our most fundamental behaviors that rely on smell. The smell of fresh dung can mean meat on the hoof. The smell of a fresh carcass can indicate predators nearby. Our ancestors were keen observers, as are today's hunter-gatherers.

Even finding a mate may be entwined in smell. The role of pheromones (chemicals released by our bodies) in sexual attraction is still unclear, but research is uncovering some of the strange ways scent alters behavior.

For example, newborns are guided to a mother's breast by scent. It turns out the mother's nipples give off odorous molecules that allow the baby to home in on its food source. And here is another nippily fact: odors given off by breast-feeding women can actually stimulate their childless female friends, although the exact "randy" chemical has yet to be identified (to the dismay of all you males).

As impressive as our sense of smell may seem, it pales in comparison to most of our faunal friends. Bears have the most impressive sniffers. Black bears have been known to hoof it eighteen miles to track down prey. Grizzlies can smell a carcass even if it's underwater. And male polar bears will trek a hundred miles if they catch a whiff of a receptive cow. That's a long way to go for some tail. . .

And here are a few other samplings from the animal kingdom: elephants can smell water from as far away as twelve miles, even if it's underground; snakes smell via their tongues, catching scent molecules and processing them via specialized organs in their mouths; and finally, the most impressive of all, the male silk moths, who can detect as few as two pheromone particles exuded from the female from six miles out!

Although our noses don't measure up to our critter counterparts, we corner the market on smelly sexual fetishes. I present to you "eproctophilia." Believe it or not, there are folks out there turned on by the smell of farts. Eproctophilia is another form of paraphilia (which we discussed last week in *Dead and Lovin' It*). It's a crazy, crazy world out there.

Smelling is part of our evolutionary heritage. It is hardwired into our brains and is one of five primary senses that allow us to navigate our world (along with sight, hearing, taste, and touch). And it is the direct association to the emotional centers of the brain that makes smell such a powerful trigger. Think about some of your favorite smells. What do they remind you of? A scene from your childhood? A long-dead relative? A favorite place? Smell can instantly transport us to a time and place far removed from the present. So, take a deep breath and lose yourself in a memory.

As for my favorite smells: a windswept beach, a wood fire at night, and bacon!!

A BLOODY HISTORY
October 25

Halloween is upon us. The sweltering summer is slowly waning, the humidity is inching downward, and the first hint of cooler weather is creeping in. Those annoying little creatures (children) have returned to school, football and hockey are underway (finally!), and pumpkins are popping up all around town.

So, to usher in the spooky season and to get everyone in the mood, we are going to explore the history of one of our most essential bodily fluids. Join me as we take a bloody journey through time.

Long before human dissection was an accepted practice, the inner workings of the body were a mystery. Can you imagine the perplexity of our ancients as they sustained bloody injuries, yet had no idea what that red fluid was, how it was made, or what purpose it served within the body? It's no wonder they ascribed spiritual and magical explanations to bodily functions.

Although the Egyptians didn't understand blood's basis, they, like many throughout history, thought the secret to curing illness was to bleed the patient. The practice of bloodletting persisted for over four thousand years and was believed to rid the body of impurities. It was also believed to restore balance of the "humors," the four elements, according to Hippocrates, that made up the body: blood, phlegm (a personal favorite), and black and yellow bile. Our first president's death was hastened thanks to this ancient treatment.

It wasn't until the dissections of Galen almost two thousand years after the Egyptians that we began to understand the body's plumbing (although he worked solely on apes). Galen recognized the fact that arteries and veins carried blood and that the blood circulated via vessels. But he mistakenly believed blood formed in the liver and passed from one side of the heart to the other. Hey, you can't win 'em all.

By the 1500s, scientists and physicians concurred that the heart was involved in circulation, although the exact mechanism still escaped them. Spaniard Michael Servetus refuted Galen's theory about the flow of blood through the heart and, although he was correct about circulation, he was later burned at the stake for criticizing the Holy Trinity.

In 1628, the British physician William Harvey correctly explained the role of the beating heart in his oh-so-popular text, *Exercitatio Anatomica de Motu Cordis et Sanguinis in Animalibus*. A real page-turner...

Thirty years later, the up-and-coming Dutchman, Jan Swammerdam, (only twenty-one at the time!) was the first to describe red blood cells. I consider his discovery one of the truly missed nomenclature opportunities in science. Think how much more colorful biology would be if our circulatory system relied on "swammerdams" to oxygenate the body!

By the late 1600s, the concept of blood transfusion was still a work in progress. The first to attempt it was Parisian Jean-Baptiste Denis, and the experiments soon spread to England. Unfortunately, they chose to transfuse their patients with the blood of animals and after multiple failed attempts, the practice was kicked to the curb where it languished for almost 150 years. Finally, by the mid-1800s, docs had worked out the kinks, mainly by sticking to species-specific blood. They found human-to-human worked best, although even transfusions between peeps had problems. Sometimes the blood would clump, and it wasn't until blood groups were identified that the mystery was finally solved.

In 1901, Karl Landsteiner identified three different types of blood: A, B, and O, and received a Nobel Prize for his efforts. A year later,

Decastello and Sturli added a fourth type—AB, and the modern era of hematology was underway.

Today, we know an awful lot about blood. It is comprised of several crucial components. Red cells contain a protein called hemoglobin, which binds with oxygen, enabling blood to transport that essential gas throughout the body. White cells fight off infection, platelets form lifesaving clots, and plasma is the fluid in which the other components are suspended. Together, these constituents make up one of the most important fluids in the body.

The average adult contains about five liters. Blood is oxygenated in the lungs, pumped out by the heart, and circulated via the arteries. Veins return the oxygen-depleted blood to the heart, where it is then pumped back into the lungs, returned to the heart, and the cycle starts all over again (although within the pulmonary circulation, the roles of the veins and arteries are reversed).

So, as Halloween draws near, take a moment to appreciate the blood coursing through your veins. Place your finger against your carotid artery and relish the beautiful beat of your heart, knowing that with every pulse, every throb, that life-sustaining liquid is delivering its fragile cargo to your hungry tissues. I'll leave you with the lovely words of Georg Büchner:

"The death clock is ticking slowly in our breast, and each drop of blood measures its time..."

BODY BANDITS
November 1

Don't tell me—you went out last night to celebrate Halloween, decked out in some silly costume, you drank too much, made a fool of yourself, and this morning, you're nursing a haunting hangover. Am I right?

Well, for those of you who misbehaved last night, I'll speak softly. For my responsible readers who dragged their kids from house to house and turned in early with a bellyful of candy, I'll squeeze the last bit of life out of the Halloween season by discussing a curious piece of ghoulish history.

In keeping with last month's scary theme, we tackled sex with the dead (*Dead and Lovin' It*), the essence of smell (*Something Smells*), and the investigation of blood (*A Bloody History*). To tie it all together, let's discuss a topic that involves each of these subjects: digging up the dead.

You don't hear much about grave robbing these days. It seems to have faded from culture, kind of like the outhouse and Lance Armstrong. Modern grave robbers are typically called "looters"— those cretins who pillage archaeological sites for artifacts and bones. I can speak for all archaeologists when I say these criminals should be pummeled. But instead of focusing on contemporary looters, let's have some fun and go back in time. Let's visit 19th century England and explore the lurid history of the body snatchers.

The grave-robbing industry was born of necessity. As medicine evolved from hocus-pocus to actual science, medical students needed cadavers on which to explore the inner workings of the body. As a former medic, I can assure you no textbook comes close to providing the up-close-and-personal experience of investigating

99

a corpse. Autopsies were part of our paramedic curriculum (see April's *Musings on an Autopsy*). And nothing beats the real thing.

But if you were an English medical student in the 1800s, there were few textbooks available. Dissection was in its infancy in this region of the world and cadavers were in short supply.

The legal system tried to remedy the problem. In 1751, they passed "The Act for Better Preventing the Horrid Crime of Murder" (even English laws sound fancy), which proclaimed that anyone convicted of murder would be executed. The murderer's corpse would then be turned over to the medical students for dissection. The law served two purposes: it deterred crime (since no one wanted to end up on the dissecting slab), while also providing fresh meat for the students.

The problem was there simply weren't enough murderers to go around. In fact, in 1831, there were over nine hundred med students in England champing at the bit for a corpse, but only eleven convicted killers. To make matters worse, few women were put to death. Between 1800 and 1832, only seven females were executed. This meant most docs graduated from medical school having never worked on a female body (unless they were lucky enough to have an accommodating spouse or sister). When the rare female was put to death, the dissection turned into a morbid free-for-all. The body was typically cut into pieces as desperate students inspected every square inch of her anatomy. Special attention was paid to the lady-bits, since unmarried students (especially the less attractive ones) rarely had the opportunity to explore a woman's nether regions.

Enter the resurrectionists. These entrepreneurs were well-versed in supply-side economics. If the med schools needed bodies, they knew just where to find them. Thus, the era of grave robbing was born.

The bandits would sneak into graveyards at night, target fresh graves, and hastily disinter the bodies. The more ambitious agents even posted scouts who stood lookout for funeral processions. The public quickly caught on, devising elaborate tombs to protect their dearly departed. Fences, locked vaults, and mortar slabs were just a few of the deterrents put in place to try to keep the resurrectionists out. Friends and families even kept vigil for the first few days, to ensure decomposition could set in, since putrid bodies were less desirable.

With grave robbing on the rise, the government intervened once again in an attempt to provide cadavers. In 1832, The Anatomy Act was passed, which not only relegated executed criminals to the dissecting table, but also included unclaimed bodies from area hospitals. These were typically the poor or destitute who died in obscurity. As for other patients, family members quickly learned to stand guard over their loved ones. If a body wasn't claimed within forty-eight hours, it was handed over to the med schools. The schools even went as far as posting "clerks," who would roam the hospital corridors, waiting and watching for a fresh corpse.

But the most famous body snatchers heralded from further north, in Edinburgh, Scotland. William Burke and William Hare went into business together, robbing graves by night and selling the bodies to anatomist Robert Knox by day. But grave robbing was hard work. They soon realized killing was easier than digging, so they abandoned their nightly excavations and turned to murder.

Their preferred method was suffocation, since it made for a tidier corpse, and they typically targeted prostitutes, since the disappearance of a hooker was unlikely to arouse suspicion. The killing spree, which became known as the "West Port Murders," tallied sixteen victims before the Williams were finally caught. Their "Burking" days were over, and, in an ironic twist of fate, Hare turned snitch and testified against Burke, who was found guilty, put to death, and subsequently dissected. Ah, sweet justice!

So, with this brief history of body snatching, our Halloween season draws to a close. Fall marches on and before we know it, we'll be celebrating every pilgrim's favorite holiday, Thanksgiving. The trick will be to devise a way to incorporate turkey anatomy into the blog. Stay tuned!

SIZE MATTERS
November 8

Think about your favorite sports. What do you like about them? Is it the speed, the precision, the intensity, the beauty? Is it a sport you enjoy playing or simply a sport you like to watch? Do you watch it for the sheer brutality (boxing) or the skimpy uniforms (women's beach volleyball)? There are numerous reasons for liking the sports we like.

One of my favorite sports to watch is hockey. Although I've never played hockey, have never even put on a pair of skates, and am positive I would end up with a serious head injury if I tried either one, I'm truly mesmerized by the game (although I abhor the fighting). Any sport in which big burly men race around, bashing into each other and chasing a puck, while performing the entire feat on skates, ranks as phenomenal in my book.

The biomechanics of hockey are astounding. The grace with which the players move, combined with the physical intensity of the sport, make for a strange and beautiful combination. It's hard to tell much about these players' builds. Their bulky uniforms disguise the bodies underneath (although I spend considerable time imagining them out of their uniforms), but it takes a highly muscular frame to perform the lightning-fast, bone-crushing maneuvers that make up this wonderful game.

Biomechanics—the forces exerted on the skeleton by muscles and gravity—are a fundamental aspect of any sport. And some sports demand a certain type of body. So, let's explore some sport-specific bodies and see how size and shape enable athletes to excel in their chosen games.

Let's start small. Professional jockeys must be tiny. Although there is no height restriction in the world of horse racing, the average jockey is less than five-and-a-half feet tall (that's about 1.7 meters for all you non-Americans). A jockey's height is restricted by his weight. The lighter the horse (including its cargo), the faster the horse, and every ounce counts. The average jockey weighs between 108 and 118 pounds (around fifty kilos). If you want to ride in the Kentucky Derby, you can't exceed 126 pounds, but that includes all your equipment (no naked bareback riding allowed). Although jockeys are remarkably small, the weight restrictions are a constant battle, especially as jockeys age. This has led to a dysfunctional culture of weight loss, where these little guys starve, sweat, and puke their way to feather weight. I'd never make it as a jockey.

What about the opposite extreme? The tallest players in the NBA stand over seven feet tall. An alligator of similar size could easily swallow a goat. Since regulation nets are ten feet off the ground, these skyscraper players, with their gangly arms, can easily dunk the ball; that is, if they're not blocked by a totem pole from the other team. The average height of a WNBA player is around six feet, although Margot Dydek, the tallest woman to play professional basketball, was 7'2". That's a lot o' woman.

There are exceptions to the rule. The shortest player in the NBA was teeny-tiny Tyrone Bogues. At just over five feet tall, "Muggsy" made up for his lack of height with lightning speed and went on to play for fourteen seasons. Miracles come in small packages.

Gymnastics is another sport that necessitates a certain frame, namely, one that's compact, flexible, and incredibly muscular. You'll never see a seven-foot-tall gymnast; there's no way to tuck that much body into a ball. Can you imagine the carnage on the uneven bars? Gymnastics is a youthful sport and puberty serves as a double-edged sword. For male gymnasts, it means increases in testosterone, which enhance the athlete's ability to perform those gravity-defying feats. For women, puberty is accompanied by an increase in body fat as the body prepares for childbearing—not good for someone whose primary job is tumbling. That's why most females hang up the leotard once they hit their twenties. But the constant wear and tear on joints and muscles limits even the men. There's no such thing as a middle-aged gymnast.

Granted, there are some sports that simply require technique, irrespective of the size or shape of the athlete. The exciting world of professional bowling comes to mind. . . Golfers are another group that falls into the "muscles optional" category (and their nerdy clothes don't help matters). I've also noticed how chunky many of those baseball players are. I guess the ability to scratch and spit has no correlation to pants size.

Bodies come in all shapes and sizes and sometimes size and shape are dictated by the sport. Want to be a lineman in the NFL? You'd better tip the scales at three hundred. Heart set on becoming the world's greatest Sumo wrestler? Better be able to consume twenty thousand calories a day. Our bodies can facilitate or impede the goals we set.

Speaking from the standpoint of a female firefighter, the physical demands are a constant challenge, and most women in the fire service must work twice as hard to achieve the same results as our male counterparts. But drive and determination make for powerful fuel, so set realistic goals and go for it. You'll never know what you're capable of until you give it a shot.

ANATOMY OF A HANGOVER
November 15

Howdy, from the hills of North Carolina, where I've escaped for a week of isolation among the lush beauty of the Smoky Mountains. Dense clouds are skirting their peaks and a grey sky is spitting snow, so I've settled before the fire to warm my cold-intolerant blood.

I've been coming here for years. The topography provides dramatic contrast to the flatness of Florida and it's nice to witness the change of seasons. The week is usually spent indulging in two of my favorites: gin and bacon (if only they made bacon-flavored gin. . . or gin-flavored bacon!).

I've been a gin drinker since my early years on the fire department. In fact, it was a fellow firefighter who introduced me to that magical libation. Gin is not intended for the novice. It creates a blissful burn as it goes down, akin to swallowing an ecstasy-laced razorblade, and the effects are intense and immediate. Fortunately, following a long apprenticeship, I am now a proficient consumer. I know just when to cut myself off before the inevitable penalty sets in: the hangover.

We've all been there. The hammering head, the nausea, the shaking, the thirst. And although you can take meds to minimize the symptoms, you simply must wait it out. It's a slow form of punishment that sets its own pace. So, let's examine just how alcohol ushers in this suite of symptoms and, the next time you reach for that fourth or fifth cocktail, you might just take heed.

Although alcohol is technically a depressant, the initial effect is a blissful lightheadedness. Alcohol's effects are based on several factors—what you are drinking, your body size, how much you have had to eat, and how fast you are drinking. A few quick shots on an empty stomach can produce intoxication in no time, especially for individuals unused to heavy consumption (aka, "lightweights").

As you drink, the alcohol enters your stomach where it is absorbed by the bloodstream and circulated throughout the body. Because drinking lowers your inhibitions, you tend to disregard the warning signs and keep on drinking. It's a vicious cycle and before you know it, you're smashed. Enter Mr. Hangover.

Alcohol wreaks havoc on your body. Even when you manage to make it home and into bed, the fun has just begun, for here come the spins. Those miserable bed spins are caused by the alcohol affecting the fluid of your inner ear. The disruption sends signals to the brain, telling it the body is moving, when in reality, you're simply hanging on for dear life, trying not to hurl. Word of warning: the spins are even worse if you add weed to the mix.

As you're busy spinning, the alcohol is toying with other bodily components. Urine output increases, which can lead to dehydration (dizziness, thirst, and lightheadedness). Your stomach lining becomes irritated, which contributes to nausea and vomiting. Blood vessels expand, causing your head to throb. And blood sugar can drop, which brings on the shakes.

On a broader scale, alcohol can trigger an inflammatory response, which your body combats via the immune system. The agents released by your immune system can cause a decrease in appetite, loss of memory, and an inability to concentrate. Alcohol also affects quality of sleep, which can intensify each of these symptoms, leaving you cranky and fatigued.

So, with all these ill effects, why do we continue to drink?? Because it's so damn fun.

Humans have been consuming alcohol for therapeutic, ceremonial, and recreational purposes for thousands of years. Evidence for alcohol dates back over nine thousand years in China's Henan Province, where folks enjoyed a "wine-and-beer-like beverage" made from fermented grapes, rice, honey, and hawthorn fruit. Using residue analysis from pottery fragments, modern concoctors were able to recreate this brew, which went on to win a gold medal at the Great American Beer Festival in 2009. Science is awesome!

And the Chinese weren't the only ones raising a glass (or vessel, I should say, since glass wasn't invented until the Bronze Age). Ancient Egyptians, Phoenicians, Turks, and Mayans were also imbibing. And wherever there has been drinking, there has been overdrinking.

So, the next time you overindulge, picture our ancient brethren in the same situation, for as long as there's been alcohol, hangovers have lurked just around the corner.

Drink wisely and stay safe! Next week I'll be writing from the great city of Chicago, where hundreds of fellow nerds and I will be gathering for the American Anthropological Association's annual meeting. I'll be sure to pack my flask. . .

HUMANS... TASTY, TASTY
November 22

It's good to be a human. After a few million years of evolution, we have finally made it. We've conquered land, sea, and space, invented medicines, phones, and computers, and inhabited every square inch of the planet (although you can still get a deal on real estate at either pole). Humans are pretty awesome.

Humans are also masterful hunters. If it weren't for hunting, we probably wouldn't be here. Scientists are just starting to appreciate the role meat eating, and subsequently, cooking, played in making humans the intelligent (for the most part) and progressive species they are today.

We tend to envision ourselves at the top of the food chain. We can haul in fish as fast as we can bait our hooks, blast our way through a herd of elk, and slay even the largest of carnivores, thanks to modern weaponry. But that doesn't mean our fellow critters don't occasionally get revenge. Let's check out some of the curious ways humans end up on the menu.

I live in the beautiful state of Florida and guess what! Our little peninsula led the way last year in shark attacks in the United States. (Suck it, Hawaii!) Of course, sharks aren't the only happy man-eaters that inhabit the Sunshine State. We boast over a million alligators which, on occasion, have been known to munch on a human. Surprisingly, the last fatality was way back in 2007 when a car thief in Miami jumped into a pond to elude the cops and was greeted by a belligerent gator. Like they say, crime doesn't pay.

There was a plethora of bear attacks in the U.S. this past year. A couple of hikers in Yellowstone National Park were foolish enough to approach a passel of grizzly cubs before momma bear gave chase and subsequently bit one of the hikers on the ass. Kinda' serves him right. If you are foolish enough to flirt with a grizzly, you deserve to lose a bit of tail. Experts say if you are confronted by an angry bear, you should curl up in a defensive ball. Seems in that position you are just asking for an ass bite.

The mountain lion (aka, cougar, puma, panther) is another carnivore that seems to fancy a human from time to time. Many attacks take place along the west coast, where lions patrol paths frequented by hikers and bikers. According to the Mountain Lion Foundation, these agile cats can leap fifteen feet into trees, jump twelve-foot fences, and reach speeds of fifty mph. What to do if attacked? Instead of rolling in a ball and getting bit on the ass, as you would in an encounter with a grizzly, they recommend confronting the feisty feline. Maintain eye contact, flap your arms to appear bigger, and make noise. Worst case scenario, chuck your granola bar in its direction. It turns out many people have avoided becoming lunch simply by mimicking a crazed chicken and fighting back.

Some animal attacks are brought on by humans themselves. These folks fall into the category of "exotic pet owner" or, as I prefer to think of them, trauma patients. These "pets" include tigers, wolves, leopards, and apes. Why people think they can domesticate a wild animal and magically suppress its natural tendency for meat (any form of meat) is beyond me.

Here's a fundamental principle of domestication: never domesticate anything that can eat you. There's a reason ancient herdsmen chose cattle, pigs, and sheep. Your chances of being mauled by a goat are pretty slim.

There is one domesticate that regularly attacks humans: the dog. In 2012, there were thirty-eight fatal attacks in the U.S. The majority of these were from pit bulls; sixty-one percent, in fact. Ironically, pit bulls account for only five percent of the U.S. dog population. According to dogbite.org, pit bulls killed 151 Americans between

2005 and 2012; about one fatality every nineteen days. But we can't fault the dogs. Pit bulls are commonly bred for aggressiveness and this, combined with their powerful jaws, makes for a lethal combination. And here's an interesting tidbit: dogs typically out-kill sharks by twenty-six percent, which means you're safer snorkeling among a school of great whites than you are walking outside to check your mail.

So, remember: even though you relish your role as master of the universe, there are still plenty of bigger and badder creatures out there that can take you down. After all, we possess the same juicy cuts of meat as our friends on the hoof and, in the eyes of a carnivore, we are all fair game. Be safe out there.

GOBBLE, GOBBLE
November 29

Thanksgiving Day has passed and if you live in the U.S. or Canada, you are probably still recovering from turkey overload. Yesterday we commemorated that mythical feast between the Pilgrims and Native Americans, where they all sat down to share in this land's bounty. (*"Would you like some smallpox with your gravy?"*)

Actually, Thanksgiving is probably my favorite holiday. Fall is a wonderful time of year and turkey day is tucked nicely between Halloween and Christmas, resulting in a blissful holiday trifecta. I've already compared the human circulatory system to a fire engine (April's *Anatomy of a Fire Truck*), but to commemorate Thanksgiving, I thought we'd have some more fun with comparative anatomy and see how much we share in common with our feathered friends. Let's begin with a glimpse at our genealogies.

Birds evolved from small carnivorous dinosaurs around 170 million years ago. This ancient ancestry accounts for the many different bird species that exist today; around ten thousand at last count. Humans, on the other hand, split from our last common ancestor (LCA, if you want to sound savvy) a mere six million years ago, give or take a million. Birds beat us on scene by a long shot.

Feathers evolved, not for flight, but probably for insulation or display. Only later were they commandeered as part of the airborne assemblage. Turkeys sport an impressive array—over thirty-five hundred. We, on the other hand, lack these colorful adornments and can only grow hair, although there are some among us who could give Sasquatch a run for his money.

Turkeys, like us, are vertebrates. Thus, we share many of the same bones, although theirs have been modified for flight. Most birds

112

have hollow bones compared to ours, which are thicker and heavier. Hollow bones make for a lighter skeleton, which is essential if you intend to get off the ground. Penguins are the exception, but their chunky little bodies have evolved for swimming, resulting in a very non-birdlike anatomy.

Like us, turkeys rely on vision over smell. In fact, turkeys can detect movement from a hundred yards out. And contrary to popular opinion, they are fairly intelligent (unlike many humans). They are keenly aware of their surroundings and can be quite friendly. Even early Europeans commented on the cordiality of the turkeys they encountered when they arrived in the New World. The birds would strut right up and cluck "howdy" just before they were clunked on the head and thrown on the fire.

Turkeys were first domesticated by the Aztecs of Central Mexico who not only bred them but also worshiped them. The ancients relied on their meat, eggs, and feathers, but also believed turkeys were the physical manifestation of one of their gods, Tescatlipoca, and held celebrations in their honor. Once the Spanish clobbered the Aztecs, they loaded their ships with turkeys and sailed back to Spain. The birds were then domesticated throughout Europe. Ironically, the Pilgrims toted the birds back to the New World aboard the Mayflower. These are some well-travelled birds.

Ben Franklin was enamored of the turkey. He referred to it as a "bird of courage" and tried to convince his fellow Founding Fathers to adopt is as the symbol for the new U.S. of A. But his contemporaries didn't share his enthusiasm and instead, nominated the eagle. Wise choice. It would be hard to kick ass around the world if your national symbol was a gobbler.

And speaking of prowess. . . Turkeys and humans also share many similarities in their courtship rituals. Their males, like ours, puff themselves up so they appear bigger and stronger. They prance around, grunting and vibrating their bodies to entice the hens. And this can go on for some time until the female finally grows bored and submits (sometimes it's just easier). The males will also mate with multiple females, if given the opportunity. Turkeys, like men, rarely turn down a chance at tail.

So, enjoy those turkey leftovers. As you munch your turkey sandwich or slurp your turkey soup, take a moment to appreciate this magnificent bird. But before you chuck the carcass, think of the long history that brought this bird to your table and how each of our bodies tells its own evolutionary tale. *Bon appétit!*

BREASTS FOR HIRE
December 6

I was cruising home from the gym the other morning, listening to the radio, when the dramatic hyperbole of the local news station broke in.

"You Won't Believe What We Found When We Investigated the Contents of Breast Milk Being Sold on the Internet!" they exclaimed in mock outrage.

What made me gasp was not their groundbreaking reporting or the fact that there were contaminants in online breast milk. What shocked me was that there are loonies out there actually purchasing bodily fluids on the internet and feeding them to their newborns! *What the Hell?!!*

So, I did some investigating of my own and it turns out breast milk is a hot item for sale on the web for parents who cannot provide their own. Granted, I don't have kids so I've never had to deal with the trials and tribulations of breast feeding and, between you and me, any child of mine would starve, since I'm built like a prepubescent tween. But I was stunned that there are people out there who would purchase breast milk from a stranger and pass it on to their young. Isn't natural selection supposed to weed out such insanity?

Not that the use of another's milk is anything new. Milk surrogates have been around for thousands of years. So, let's spend a moment reflecting on this cultural practice, but let's begin with a quick overview of the female anatomy. We have already covered the nether regions (October's *Below the Equator*) so let's head north. Welcome to "Breasts 101."

115

Humans are mammals, meaning we are hairy, warm-blooded milk producers who nurse our young. Breasts are the means by which humans nurse, although their role in reproduction has been vastly overshadowed by their use in entertainment. Each breast is a complex structure of fat and connective tissue, lobes and lobules, ducts and nodes. All of this is arranged (somewhat) concentrically around a nipple, formally known as the areola.

Women produce milk in specialized cells called alveoli and the milk is carried via ducts to the nipples, where it is then consumed by the hungry newborn. What I didn't know is that all women have about the same number of alveoli, regardless of breast size, so perhaps my kids wouldn't starve after all. But what happens when a woman can't produce milk? Or, worst case scenario, the mother dies in childbirth? Enter the wet-nurse.

The practice of wet-nursing goes back over four thousand years. Back then, if a woman failed to lactate, there were several remedies she could try before resorting to a surrogate: having her back rubbed with an oily concoction of fish bones, eating certain types of fragrant breads, or rubbing her malfunctioning breasts with poppy plants. If all else failed, a wet-nurse was brought in.

Wet-nurses were typically members of the working class, if not outright slaves. Selection of the nurse would sometimes depend on the quality of her milk, which was gauged by the "fingernail test." A drop of her milk was placed on the fingernail and if the milk slid off, it was considered too watery. If it clung to the nail when turned upside down, it was considered too thick. The ideal nurse produced milk that was "just right!"

In ancient Greece, wet-nurses were not only utilized out of necessity, but were the preferred choice of higher status individuals. This trend carried on into the Renaissance, where aristocratic women were too busy playing cards and attending the theatre to worry about nursing. Besides, it was believed breastfeeding would ruin the figure and, in fashion-conscious France, style always trumped suckling.

But as the practice spread, so did its opponents. Many believed the use of a wet-nurse would harm the child by passing on physical or psychological defects from the nurse, especially if the nurse were a redhead! Redheads were believed to be ill tempered—a trait that tainted their milk. Others believed breast milk contained magical qualities meant to be shared between mother and child; thus, it was a mother's "saintly duty" to nurse her child.

Today, women have more options. Gals who cannot, or choose not, to breastfeed commonly use formula, and there is a plethora to choose from. Although formula lacks many of the benefits of breastfeeding, especially when it comes to immune response, it is still a practical option.

But should a mother insist on breast milk for her child, there are ways to obtain it without resorting to Craig's List. The Human Milk Banking Association of North America (HMBANA) provides outlets for safe, pasteurized donor milk, as well as donation portals for women who have milk to spare.

So, play it safe and use good judgment in your hunt for milk. After all, you wouldn't accept a blood transfusion from some dude on the web.

Trust me—your baby will thank you.

GOOGLING THE BODY
December 13

I have always been curious about the body; my body, the bodies of my friends and families, even strangers' bodies. Whether I'm at the gym, presenting a lecture, or simply in a crowded place, I can always amuse myself by checking out the bodies around me. We all do it. "People watching" is a most enjoyable pastime.

Although we are all one species, the range of dimensions and proportions of the human body are truly astounding. We've already discussed how size and shape relate to sports (October's *Size Matters*), but this week I want to share a little-known fact about my blog.

I have been writing the blog now for almost nine months and ironically, much like a pregnancy, it has been a period of intense development—for me as a writer and for you as readers. I have watched my audience grow in leaps and bounds, reaching *almost* every continent on earth (will someone *please* share the blog with their buds in Antarctica!!).

What you may not know is that I monitor my readership through statistics provided via Blogger. I can track the number of page views, where in the world these hits occur, and—most interestingly—the Google searches that lead you to my page. Don't panic! I can see the search terms but not who's searching!!

So, I thought it would be fun to share with you some of the bizarre things you folks are searching for out on this vast frontier that is the Internet. Join me as we Google the body.

Being a science geek, the task of analyzing subject categories and crunching numbers left me all atingle. I was able to separate the search terms into five general categories:

1. Teeth
2. Bones
3. Blog Title/Website
4. Sex (of course!)
5. Oddities (and there are some strange ones in the mix)

People have a lot of questions about teeth and bones, which isn't surprising. With over fifty teeth (baby plus permanent) and over two hundred bones in the average body, it makes for a lot of complexity and thus, many questions. "Childs mouth teeth x-ray" and "8 years old child teeth x-ray" are perhaps parents concerned with their child's oral health or a researcher in need of a graphic (since I try to hold your attention via cool pics in the blog). There are also those interested in culture, like "8 teeth grillz," although when I Googled this term, I came up with a YouTube video about ridding your dog of bad breath. I still don't understand that one.

The bone searches come in a wide variety. "Bones of the face," "appendicular skeleton," "axial bones," and "cranial sutures" were obviously folks on an anatomical info quest. "Bones of the head quiz" was perhaps someone searching for a bit of nerdy fun.

Many find the blog via post titles or my name. "Anatomy of a fire truck" and "anatomy of hangover" are perfect examples. These are probably individuals who have attended one of my lectures or heard about the blog via word of mouth. So, keep those coming!

It's no shocker that many of you are cruising the Web in search of sex; therefore, it's no surprise that my most popular post was September's *A Natural History of the Penis* (you boys are obsessed!). But many who find my blog are trying to puzzle out the complexities of sex, and these complexities come in a range of issues.

There are the curious, such as "females private parts" and the ever-present "penis length." There are those who have had mishaps, like "anal scarring" or (my all-time favorite) "women's private parts that stink." And then there are those that move me, the searches that stem from a concern or insecurity, such as "teen penis growth." Searches like this speak of our fears and worries; our feelings of inadequacy that flare especially during puberty (don't worry, little buddy. . . it'll get bigger!).

And finally, there are the truly strange searches that somehow land on my doorstep (although this says something about my blog's content). For some reason there are a lot of questions about Chinese foot binding, which have found me via May's *Beauty of Feet*. It seems I've also cornered the market on leeches. "Leech anatomy," "leeches," and "2013 leech saliva" are just a few examples of those who stumbled onto the blog via May's *Ode to the Leech*. And then there are the searches I don't even understand. Someone will have to explain to me what the hell "feet transfusion cartoon" means. . .

The Internet is a powerful tool—not only for providing information, but for bringing people together. It thrills me to see this little blog being read by someone in Yemen, or the Sudan, or Thailand. To know that someone in Afghanistan, France, Malaysia, or Ukraine might be learning, or at least chuckling, because of something I wrote is incredibly rewarding.

So, keep reading and keep sharing, and I'll try each week to bring a little bit of info your way. Thanks to you all!

A MOTHER'S TOUCH
December 20

With Christmas just around the corner, 'tis the season for family gatherings. This provides either a blissful rush of joy or a pang of fear and dread, depending on your family dynamics. The holidays bring us together, whether we like it or not.

Family traditions are big around the holidays. I always flash back to the customs my mother instilled: tasteful decorations, plenty of homemade treats, and a *mandatory* live tree. The one year we tried out a fake, I cried and cried.

Mothers have a tough job. There's nothing more demanding than a newborn and that's just the start of decades' worth of nurturing. I was lucky. My mother provided a loving childhood for me and my siblings and her life lessons still reverberate when I think back to all she taught me.

So, stop and reflect for a moment on all that mothers do. For one thing, they signed on to giving birth to you and, let's face it, pregnancy ain't no picnic.

From the moment of conception, the woman's body begins to change. Even before the pee strip turns blue, hormones inside her body are gearing up for the ordeal. Rising levels of estrogen and progesterone usher in the traditional "morning sickness." The breasts swell (no complaints from the menfolk), urination increases, and fatigue latches on for a nine-month ride. Blood vessels dilate, leaving her hypotensive and dizzy; food is absorbed more slowly, causing heartburn and constipation; and her emotions climb aboard the pregnancy rollercoaster. Yes, it's a joyful experience. . . and this is only the first trimester.

It's hard for me to fathom my mother going through this four times, especially since I'm too gutless to even give it a one-time shot. But she came through like a champ and even managed to guide four kids to adulthood.

I've already discussed my haphazard childhood in April's *Disfigured*. My injury calamities put my parents through the wringer, but mom was pretty savvy when it came to managing our maladies. She was stern when it came to illness (German roots). Whenever we tried to dodge school, she would nail us with her standard response: "*You can make it!*" It became a joke around our house. I could have lost a limb in a lawnmower mishap and she would have scooted me right out the door with my book bag, lunchbox, and a trauma dressing.

She was a master at tending wounds, and with four kids, she practically ran her own M.A.S.H. unit. Our bathroom was fully stocked with Band-Aids, Neosporin, and baby aspirin and she administered each with the practiced precision of a drill sergeant. I remember her driving me to the hospital when I broke my arm, applying cold compresses when I knocked out my front teeth, and tenderly nursing me through chicken pox, measles, and a steady onslaught of ear infections (although according to my grandmother, she pushed my bassinet into the kitchen and shut the door after a particularly long night of my wailing).

She was also a patient instructor when it came to the mysteries of sex. I remember her gently explaining my baby brother's bizarre genitalia (although because of her I still refer to testicles as "plump-plumps"). She explained the nourishment of the fetus when I pointed to his gross little umbilical stump, and she guided me and my sisters through the perils of menstruation with hands-on lessons in feminine protection.

So, as you gather for the holidays, take a moment to appreciate your mother. Put aside any petty disagreements over clothing styles, makeup, or your choice of a spouse. Practice patience when she dictates how long the turkey should roast or points out the lumps in your gravy. And keep in mind how priorities and perspectives shift as we age.

My mother died of cancer on Christmas Day when I was twenty-three. Since then, the joy of the holiday is always laced with the pale taint of grief, the faint echo of loss. But I focus on the wonderful traditions she instilled, the warmth and happiness she so generously spread, and the valuable lessons she taught through her ever-patient instruction. These things live on.

Happy Holidays.

BIGHEADS
December 27

The other day, I sat down to enjoy a giant grapefruit my coworker plucked from his accommodating tree. That's the advantage of living in Florida: free fruit. As I admired the grapefruit's beauty and heft, it struck me that it was the approximate size of a newborn's skull. I filed the information away, sliced into the fruit's meaty center, and feasted on its contents.

But it got me thinking about head size. *Homo sapiens* are a big-headed bunch. How we acquired such gargantuan gourds is a long and convoluted evolutionary tale, so I'll stick to the highlights.

Natural selection is all about "advantage." If a mutation bobs to the surface of the gene pool and happens to confer an advantage— whether it's the ability to run a bit faster to elude a predator or a knack for exploiting a new food source—chances are he or she will leave more offspring and pass on that advantage. Through time and successful breeding, the advantage may become a staple, otherwise known as an adaptation. And the better adapted one is to his environment, the better his chances of survival.

So how did our heads get so big? Our skulls, which serve as protective shell and handy carrying case, form atop the underlying tissues, therefore, we can blame our bigger brains.

Having a bigger brain is definitely an advantage. The more neurons you possess, the more complex the organ. But it's not just overall size that matters. The brain of a walrus is about the same size as ours, but you won't catch a pinniped performing calculus anytime soon.

It's the size of the brain in relation to body mass that really counts— what scientists refer to as the encephalization quotient (EQ). To give you an idea of EQs, take our hooved friend, the horse. Horses are pretty smart. They usually follow directions and can be trained to perform a number of nifty tricks. The typical equine has an EQ of around 0.8. Two of our other clever domesticates, dogs and cats, have EQs of 1.1 and 1.0, respectively. Humans, on the other hand, have EQs of around 7! We are the true brainiacs.

We are still trying to tease out exactly when this advanced wiring emerged among our ancestors. By comparing skulls of our various kin, we can track brain changes over time. The neocortex, the outer region of the brain responsible for conscious thought, expanded by the time archaic humans emerged on the scene (around five hundred thousand years ago) and the temporal lobes, those regions on either side, are twenty percent bigger in modern humans compared to our predecessors. That's important, since these lobes help us organize memories, aid in learning, and allow us to store information—all vital skills associated with the development of culture.

Our modern brains contain over a billion neurons and it's this complex wiring that enables us to perform many intellectual feats that elude other animals. Reasoning, problem solving, forethought, and language are just a few of the impressive abilities made possible by our large brains (although some animals possess some of these skills to a limited extent). And it was the fine-tuning of these abilities that enabled *Homo sapiens* to invent things like complex societies and technology. You'd be hard-pressed to design a computer if you possessed the neurological complexity of a squirrel.

But our big brains come with a hefty price tag. For starters, they are metabolically demanding. About a third of the energy you produce each day goes to fueling that giant melon and without a constant influx of sugar and oxygen, it quickly dies. That's why immediate CPR is so critical in sudden cardiac arrest. Ventilating the patient and performing chest compressions not only provide oxygen, but circulate it to starving tissues. And the brain is at the top of the list, for without the brain, the rest of the body isn't much use.

Another problem with a big brain is squeezing it through the birth canal. For this we can blame our mode of transport. Bipedalism (walking on two legs) places certain architectural demands on the pelvis. Our legs must be aligned below our trunks for efficient walking and running. If not, we'd walk like a crocodile, which can sprint for short distances, but will never win a marathon.

But a woman's pelvis can only flex so far before things go wrong. That's what can make childbirth such a dangerous endeavor. Hemorrhage is a common cause of maternal death, which is understandable when you consider the size of the newborn's head in relation to the dimensions of the birth canal. Natural selection compensated by limiting pregnancy in humans to nine months. This way, the baby emerges before its head becomes too big to pass. The drawback: a defenseless newborn with an undeveloped brain, who is completely dependent on others for survival. Let's face it—baby humans are basically helpless little blobs that can't even lift their cumbersome noggins. Pitiful.

In closing, every physical attribute is a tradeoff. Bigger brains may make childbirth more problematic and gobble up much of our energy, but they enable us to do some amazing things. As I bang out this blog on my computer, I'm surrounded by the evidence of human ingenuity—all made possible by a big brain. The clothes I wear, the car I drive, the house I reside in, not to mention my electricity, phone, and medicines—none of these things would be possible without those demanding organs that sit atop our shoulders. So, hurray for the bigheads!

HAIRY BEASTS
January 3

Over a quiet cup of coffee the other morning, I got to thinking about naked men (happens more than I care to admit). It occurred to me that, throughout my lifetime, I've seen more than my share of naked males. I'm not talking the Internet or magazines, I'm talkin' live nudes. I attribute my exposure (or should I say, theirs) to two things: first, my thirteen years as a medic; second, lifestyle.

As a paramedic working some of Orlando's less reputable neighborhoods, I encountered a lot of trauma. Shootings and stabbings were regular occurrences and the first rule in trauma care: expose the wound. When you're dealing with multisystem trauma—say a bullet-riddled drug dealer—your first step is to strip 'em down. Paramedics are wizards with scissors, and since most of my trauma victims were males, I gazed upon many a naked dude.

As for my lifestyle. . . an irrational fear of commitment has caused me to skirt marriage most of my life, thus I've had the freedom to do a bit of comparison shopping. And what I've found is the male body not only comes in a vast array of shapes and sizes, but it also presents an infinite range of hairiness. Let's discuss.

Why do we have body hair in the first place? If we were reptiles, we'd be decked out in scales, dermal plates, or leathery skin. However, we are mammals, aka, hairy milk producers, thus we rely on body hair as a form of insulation. Although we appear to have much less hair than our fellow primates, it's really an illusion. We have about the same amount as apes (about a million hairs, give or take), only ours are much finer. Research suggests human hair was minimized via natural selection to avoid parasites (fleas, lice, and ticks, to name a few). Sexual selection may have also played a role, with less hairy individuals being preferred over their furry counterparts.

Hair forms within specialized follicles of the epidermis. As new cells are produced within the follicle, older cells die and are pushed out. The cells harden as they exit and blend with a protein called keratin. The result? A strand of hair. Scientists are still teasing apart the genetics that control hair characteristics, but texture and thickness depend on the size of the follicles and the density of the shaft. Who knew dermatology could sound so sexy?

Men have hair in the strangest of places. I'd never imagined noses and ears could sprout such foliage. Although those hairs serve a purpose—filtering out dust and particles—as men age, changes in hormones promote unwanted hair growth. Recommendation: invest in a nice set of clippers.

And what about chest hair? Some women love it and like nothing more than to run their fingers through a heavy mat. Chest hair, like pubic hair, is part of the testosterone-driven body changes that accompany puberty. Perhaps a hairy chest merely gives the female something to hang on to. Kinda' like a horse's mane.

And speaking of pubic hair . . . there's been much debate over why we have it. Science suggests pubic hair, like underarm hair, is for trapping pheromones. Pheromones, which appear to play a role in sexual attraction, are released by the body and blend with bacteria decomposed by secretions from your sebaceous glands. Each person produces their own aroma, based on their MHC (major histocompatibility complex, which plays a role in immune response). This heady mix of odors can infuse your armpits and crotch, thus attracting a partner—at least in theory. In some cases, all that hair simply results in a funky stench. In that case, hit the showers.

I'm also curious about a hairy phenomenon I've witnessed along our Florida beaches. Why is it the hairier the man, the smaller the Speedo? Is it some unwritten bathing suit creed? If you're going to sport a banana-hammock, at least have your significant other run a razor over your torso before you hit the sand. The public will thank you.

And finally, we must address the balding issue. I imagine men feel about balding the way women feel about cellulite: it's one of the harsh realities of an aging body. Yes, some men come through unscathed, and they have their genetics to thank for it. But for many men, male pattern baldness is a fact of life. But here's a bit of hope: bald is in! So, shun the comb-over and reach for the razor. We gals prefer smooth skin over a hair flap any day.

Fortunately, men are now taking a more active role in body hair maintenance. The era of "manscaping" is upon us and, in my opinion, is long overdue. For centuries, women have plucked, shaved, and waxed their way to baby smoothness. It's about time you men joined in on the fun. So, trim those orifices and wax that back and perhaps we'll forgive you for leaving the toilet seat up.

KILLER CULTURE
January 10

Two weeks ago, we discussed the advantage of a bigger brain (December's *Bigheads*). Even though it comes with a few drawbacks, like getting stuck in the birth canal and consuming a large chunk of our energy, it enables us to do some pretty amazing things. And as our giant melons evolved, they ushered in one of our most fundamental aspects of humanness: culture.

Culture, in its most simplistic definition, is a suite of behaviors that are transmitted within a population. In reality, culture defines us: the things we believe, the technology we use, the social customs we follow, even the way we behave. All these things are guided, if not dictated, by the culture in which we live.

The best way to appreciate your culture is to venture into someone else's. I've had the good fortune of working and studying abroad, and the experience of living in a different culture, even for just a few weeks, makes you appreciate just how culture-bound we humans are. I trained in London and even the Brits, who are culturally similar to us, have their own odd customs (they sure are fond of the word "toilet"). Then again, I'm sure many aspects of Americana send them reeling.

The development of culture has allowed us entry into some of the most inhospitable climates on earth. Clothing and technology keep us warm as we scale Everest, cool as we traverse the Kalahari, and enable us to exploit just about any ecosystem on the planet. But just like the drawbacks of having a big brain, culture has a way of biting back. Let's take a peek at a few of the ways culture kills.

Cooking revolutionized how and what we eat. It made foods more nutritious and palatable, and allowed us to redirect energy once used for digestion to fuel our big brains. But humans aren't programmed to live in a world of perpetual abundance. Our hunter-gatherer ancestors would stock up when good fortune befell them—say the taking down of a juicy mastodon—but our bodies are ill-equipped for round-the-clock takeout and all-you-can-eat buffets. The result: dramatic rates of obesity plaguing much of the developed world.

Our methods of cooking also cause problems. What began with roasting meat and veggies over an open fire has morphed into calorie-laden, artery-clogging techniques that transform our food into delicious but deadly morsels. Is there anything finer than a warm donut, fresh from the fryer? Or a succulent piece of fried chicken accompanied by a heaping pile of mac and cheese? (Pause while I wipe the drool from my computer). Human ingenuity has come up with some pretty devastating methods of food preparation that, I admit, make for some scrumptious delights, but that take a radical toll on our health.

Transportation also plays a pivotal role in our culture and has transformed the capacity and expediency of travel. Plaines, trains, and automobiles, not to mention boats, ships, and rockets, enable us to go farther faster than ever before. But these people-movers are also people-killers. According to the National Highway Safety Traffic Administration, 2011 saw some twenty-six thousand people die in vehicular crashes. The good news: deaths have trended downward since 1990, I'm guessing due to stricter seatbelt and drunk-driving laws.

Transportation via the skies enables us to hop from continent to continent, but if you are unlucky enough to be on a jet that plunges to earth, your chances are pretty slim. In 2012, there were a total of twenty-three commercial airline crashes, killing 475 people. And this doesn't include all the smaller crafts that fell from the sky. Fortunately, developments in technology, such as navigation, early warning, and weather prediction, have improved the safety of air travel, making 2012 the safest year on record since 1945.

Sports have been a part of culture going back millennia. Whether it's the naked gladiators of Rome, thrashing it out on the sawdust floor of the Coliseum, or the ancient Maya playing ball in the jungles of Central America, sports are fundamental to human nature and stir something primal within. But they also take a toll in the form of sports-related deaths. Traumatic injuries from high-speed crashes, head injuries from contact sports, and full-body splats by sky divers are just a sampling of the many ways humans die for their sports. And still, we play on.

And finally, that most deadly aspect of culture: war. Battles fought over religion, politics, and ethnicity are basically culture wars magnified to horrific scales. And as long as there has been culture, there has been conflict.

Just as parts of our anatomy are trade-offs, culture also comes at a price. It is a powerful tool that can enhance and prolong life, but it can also cause us to behave in ways that are counterintuitive to survival. Fortunately, for every drawback, every deadly action, there can be a cultural reaction. Peace accords, safety technology, and education are just a few of the ways culture compensates for our lack of tolerance, judgment, and compassion. We just have to keep prodding it in the right direction.

A TALE OF TWO CONDOMS
January 17

It was the best of times, it was the worst of times, it was the age of wisdom, it was the age of foolishness... Thus, Dickens sets the stage for our history of that most essential intercourse accessory: the condom.

The history of the condom forms a dual narrative, one divided by Charles Goodyear's 1839 discovery. His ingenuity would transform the condom from primitive to proficient by creating a material with endless application: latex.

But I've gotten ahead of my story. Let us go back in time to a simpler era, when men were men and sheep were afraid.

No one really knows exactly when or where the condom was first invented, but it pops up (forgive the pun) throughout history in many parts of the world. A cave painting dating to over seventeen thousand years ago shows a man with a sheathed penis, but you boys have been adorning your members for millennia. I'm curious whether it's truly a prophylactic or simply a kitschy decoration, since penile décor shows up in literature throughout the ancient world. Sheaths were also worn for protection in battle (ouch!), to prevent insect bites (yikes!), or to ward off evil spirits (boo!). They could be worn to represent rank or as decorations to promote fertility, since nothing says romance like a bedazzled Johnson.

The earliest description of the condom was by anatomist Gabriello Fallopio (of Fallopian tube fame) in 1564. The actual word "condom" first appears in a 1706 poem, but its origin remains elusive. And just as men have concocted numerous names for their organs, condoms also sport a range of appellations.

Germans refer to them as "Fromms," after the manufacturer who sold over fifty million a year before abandoning his factory to flee the Nazis. Our friends in Germany also use the term "naughty bags," when the mood strikes them (I can just imagine Hitler requesting one.). Other names include "bullet proof vest" (Hong Kong), "safety tool" (Hungary), and "penis hat" (Nigeria). Ironically, the French and English have assigned each other's names to their condoms—the English call them "French letters" while the French repay the favor with "English caps." Perhaps the terms are meant to imply that, should you shun protection, you'll end up like those "other" guys with a raging case of syphilis.

In the pre-latex world, men had to be creative in their choice of materials. Bladders and membranes from sheep and goats were the common choice. Even into the 1700s, condom makers would partner with local butchers, buying up innards that would be cleaned, cured with sulfur, dried, and then molded into shape. The resulting products were nothing like their modern counterparts. They were expensive, many contained holes and, worst of all, they were reusable! That's enough to drive even a randy pubescent to abstinence.

Salve-coated cloth was also used, and these little bonnets were secured to the penis by a piece of string. The more daring shunned the full sheath and simply wore a small "skull cap," although I can't imagine how they kept those little guys on without cutting off blood flow.

But in 1839, the curious Charles Goodyear stumbled onto the discovery of vulcanized rubber and the era of latex was born. His crafty chemistry used sulfur and lead oxide to alter rubber's molecular bonds, making it stronger and more elastic than its natural counterpart. Latex would transform condom construction and not only prevent a slew of unwanted pregnancies, but also revolutionize the fight against STDs.

Today, there are basically two types of condoms: latex and lambskin. According to the FDA, always choose latex, since lambskin (aka, natural) condoms may not protect against HIV, hepatitis, or the herpes viruses. According to the CDC, studies have shown that latex condoms "provide an essentially impermeable barrier to particles the size of STD pathogens" and their use has dramatically reduced incidences of HIV in places like Africa, Southeast Asia, and South America. If the package doesn't say, "For the prevention of sexually transmitted disease," keep shopping!

Alas, some folks aren't interested in disease prevention when choosing their condoms. There is a whole slew of novelty condoms on the market for those seeking a kicky way to amuse their partner. Some of them even light up and play music! With all that stimulation, you may not even need a partner.

So, the next time you glove up, think about the history and invention that went into that little rubber sheath and just be glad the days of goat bladders are behind us. Be safe!

FECAL FOES
January 24

Here's a question: When you flush your toilet, what goes through your mind? Do you ever think about the destination of your deposit? Ever marvel at the human ingenuity that makes your excrement magically disappear?

I do, especially when I stay in a large hotel. I think about all those rooms, with all those toilets, filled with all that poo, all flushing into some mystical repository. It boggles my mind.

Those of us in the developed world take sanitation for granted. We seldom think about the days of yore, when sewage ran in the streets and turds abounded in our waterways. Yes, those were the days.

The sanitation conundrum grew in concert with social complexity. As populations grew, so did issues of waste removal. As people clustered in towns and cities, they could no longer rely on the simple measures of times past (squatting in the woods or digging a latrine). Imagine, for a moment, your neighborhood, minus the sewer system... pretty scary, huh?

Cesspools were the earliest form of repositories. They could be small (underneath a house) or large (serving sections of a town or city) and merely served as dumping grounds (no pun intended) for routed waste. But over time, they would fill up, leak into nearby wells, and emit rather foul odors. The waste needed to be channeled; preferably someplace far, far away.

The Romans were pioneers when it came to sewers. As their cities grew, the age-old practice of tossing your waste into the streets became obsolete. So around two thousand years ago, they devised their first sewers. These sewers serviced public baths and latrines where folks would gather to bathe and poo (or preferably, poo and then bathe). Water flowed in channels underneath the latrines, swishing waste away from the city and (unfortunately) into nearby rivers.

Londoners followed suit, although much later in history. By the 1800s, overflowing cesspools became such a problem that it was decreed lawful to empty them into the nearby Thames (even though it served as the major source of drinking water). But it took an unusually steamy summer in 1858 for Queen Victoria to take action. The "Great Stink," as it so fondly became known, was just the impetus for the construction of a new sewer system. The Queen even constructed an underground railway from which an excited public could cheer as they dedicated their new sewer.

Cities around the world latched on to the new craze of poo-free streets. Parisians are so proud of their early system that they have a museum honoring its glorious and malodorous history.

Not only was sanitation a matter of simple decency, but it was also a major health issue. Some of our scariest diseases depend on the "fecal train" for delivery to their next victim. And the worst scenario is when our mouths serve as depot. The "oral-fecal" route is the common mode of transport for many pathogens. If you think you aren't susceptible, think how many times you put your hands to your mouth throughout the course of the day. And it doesn't necessarily have to be *your* feces. All it takes is that gloveless cook forgetting to wash his hands before constructing your mouth-watering burger for the fecal train to unload its cargo.

And the list of potential pathogens is long and frightening: hepatitis (the A and E varieties), typhoid, and cholera, not to mention shigellosis, certain viruses (rota- and entero-), and our good friend, E. coli. And contaminated food or a dirty handshake are not the only ways these critters make it into our mouths.

Keep in mind there are several sexual practices that may bring you face-to-face (literally) with feces. Anal sex is an obvious method. Once you go this route, be sure to switch the condom before switching to oral. Another culprit is anilingus—commonly referred to as "rimming." Without getting graphic, I'll simply describe it as "involving the tongue and anus." You'll have to use your imagination. And possibly the most at-risk individuals are those rare "coprophiliacs." Yes, there are individuals out there who actually get sexually aroused by poo. Whatever floats your boat. . .

Although you may not give it a second thought, flushing your toilet should be accompanied by Beethoven's Hallelujah Chorus. It is truly a wonder of human ingenuity that, when functioning properly, keeps us safe from scads of nasty bugs.

According to the World Health Organization, by 2015 there will be about 2.7 billion people without access to basic sanitation. Those of us with toilets are certainly the lucky ones.

QUACK! QUACK!
January 31

A while back, I was presenting a lecture at the Orlando Public Library and I'd arrived early, so that I had time to peruse their used bookstore. I unearthed a gem: *Doctors of the Old West*, published in 1967 by Bonanza Books. To me, the only thing more fascinating than medicine is outdated medicine, so I plunked down on a stool and tore into it.

The author, a Mr. Robert F. Karolevitz, sets the stage with an exposé on Native American medicine, which he describes in all its gory detail (he provides no references, so you must wonder if these aren't simply urban legends). He opens with a juicy anecdote:

With permission of his tribal chief, the Apache medicine man placed the two babies back-to-back, and with a single bullet, killed both of them.

What an icebreaker. Bob goes on to explain that the infants were dying of smallpox anyway, and that by killing both infants with one shot, the shaman would only be credited with one death. How economical.

To his credit, Mr. Karolevitz does expound on the many natural remedies utilized by the "gourd-rattling incanters," as he so graciously refers to them, and credit is certainly due. Native Americans had extensive knowledge of their natural world and relied on herbal remedies to treat their ill and injured. Grape and elderberry were blended into tonics; poultices of skunk cabbage and honeysuckle vine were applied to sores; and teas from a number of plants, such as sagebrush and willow, were used for diarrhea and upset stomachs. Since plants represent the earliest forms of medicine, going back over five thousand years in China, it's no wonder the Indians were working wonders with weeds.

Robert then describes the miraculous changes that took place once "civilized" medicine arrived on the frontier (his quotations, not mine). This so-called civilized medicine showed up just in time to treat the natives still struggling against the onslaught of diseases toted aboard the *Nina*, the *Pinta*, and the *Santa Maria*, along with other pestilence-ridden European ships.

But our civilized medicine has boasted quite a few "gourd-rattling incanters" of its own. They may not have literally shaken gourds, but they might as well have, for some of their early remedies certainly didn't do anything for the patient, aside from expediting death.

The Merriam-Webster dictionary defines "quackery" as "the methods and treatments used by unskillful doctors or people who pretend to be doctors." And these so-called "quacks" have a long and colorful history. Let's take a look.

One form of quackery that became part of pioneer life in America were the "snake oil" salesmen, the most famous being Clark Stanley. Mr. Stanley, aka, "The Rattlesnake King," drew large crowds by throttling rattlesnakes while he pitched his "medicine," which was supposed to heal everything from toothaches to broken bones. Stanley's potion, like so many concoctions being sold across the country, wasn't medicine at all. His snake oil turned out to be mineral oil mixed with a bit of beef fat. A dash of turpentine gave it that authentic "mediciny" flavor and, ironically, when Clark was forced to fess up, he tried shifting the blame, attributing his potion to an Indian medicine man. Quack!

Another form of quackery relied on the shape of one's head. Dr. Joseph Gall believed that a person's moral and intellectual abilities were based on the size and shape of his brain. Since the skull conforms to the brain, he believed trained practitioners could simply "decode" one's personality by translating the bumps on the skull. Dr. Gall demonstrated his technique, known in professional circles as phrenology, by identifying certain nodules on the heads of criminals. He also identified bumps associated with courage, cleverness, and murderous instincts. Soon, phrenology parlors were cropping up all over town, where one could go to have his head read and his personality deciphered. I wonder if Gall ever identified a bullshit bump? Quack! Quack!

But the most outrageous and "ballsy" form of quackery goes to Dr. John Brinkley, who wasn't really a doctor, but paid five hundred bucks for a fake diploma before launching the most preposterous transplant scheme in the history of medicine.

When a male patient of his complained about a lack of sex drive, the good doctor came up with the perfect solution. Brinkley's previous position as house doctor at the Swift meatpacking company had exposed him to the enthusiastic mating activities of goats, so it made perfect sense to implant those hypersexual goat testicles into his flaccid patient. The new-and-improved patient was able to miraculously impregnate his wife, and before Brinkley knew it, business was booming. Alas, after performing over sixteen thousand testicular transplants, his medical "license" was revoked. He did, however, die a very wealthy man. Quack! Quack! Quack!

Even in this day of modern medicine, quackery still abounds, and the Internet provides a most expedient means of spreading it far and wide. So, beware of anyone claiming to reverse the aging process, cure your baldness, or magically grow your penis. If it sounds too good to be true, it probably is.

THE DEBRIS OF LIFE
February 7

Each of us experiences insomnia at some point in our lives. I tend to wake around 3 a.m. on a regular basis and once my brain starts churning, it usually takes me at least an hour to downshift before sleep returns.

During one of my recent 3 a.m. planning sessions, it struck me that all the plastic garbage bags I have ever used are stuck somewhere in a landfill, just sitting there. Think about it: every garbage bag you've ever stuffed, taken out, and placed on the curb is probably lying dormant amidst all the other refuse from all those other people.

It got me thinking about the waste our bodies produce. I'm not talking about poo (since we covered that two weeks ago in *Fecal Foes*). I'm talking about all the other elements cast off from the living, breathing bodies we occupy.

I did a little math. Now, bear with me; math is not my forte, but I kept it simple. The average human breathes about twenty times per minute (give or take a few breaths, depending on your age and size). That's twelve hundred breaths per hour, over twenty-eight thousand per day, and over ten million each year. And that's if you're just sitting in the recliner, watching a hockey game (although those players tend to make me breathe a bit harder). On average we expel about two hundred milliliters of CO_2 per minute, 288 liters per day, and over one hundred thousand liters per year. That's a lot of hot air.

And what about our hair? Mine's boy-short, so it's not a major issue, but loss is a normal part of the life cycle of our locks. About ten percent of the hair on our heads is in a "resting phase," which means it has stopped growing and is just lying there. After a couple of months, it falls out and new hair grows in its place. We humans lose about one hundred hairs per day. That's over three thousand per year, which may be on the low end, depending on your age and genetics. That's a hairball at least the size of a cantaloupe.

Then, there are our nails. Nails are made of the protein keratin and are not only affected by age, but by health. Good nutrition promotes healthy nails. Poor nutrition, along with certain drugs (chemotherapy) can hamper growth and even cause nails to fall off. Fingernails grow at the rate of three millimeters a month (that's around a tenth of an inch, for us Americans allergic to the metric system). Growth is faster when you are young. It's also faster if you're pregnant. Can you image the nails on a pregnant teen?? Yikes! It is also a scientific fact that your fingernails grow faster than your toenails. It takes about four to six months to regrow a fingernail from scratch. It will take you a year or so to regrow that toenail.

What about urine? According to the folks at Harvard Medical School, the average person produces about six cups of pee a day, but this is highly dependent upon how much you drink. Three or four cups of coffee in the morning will send you running to the john more frequently. Frequency also depends on the size of your bladder. A teeny-tiny person probably sports a teeny-tiny bladder. They can't compare to, say, someone in the NBA. I bet their bladders are the size of grapefruits.

And here's a little urinary side note: if you are well hydrated, your urine will be a pale yellow. The darker the shade, the higher the concentration. Certain foods can alter the color. Vitamins can make it bright yellow; carrots can turn it "sunset" orange. If you are a regular consumer of asparagus, you know it not only tints urine a greenish hue, but it also gives it a funky smell. And should you ever pee red, please see a doctor.

Ironically, urine provides a proper segue into our final topic: cells. The yellow color of our urine is due in large part to the shedding of old cells. And boy, do we shed some cells. The lifespan of a cell depends on its type. Take our blood, for instance. Red blood cells live for about four months. Not so for white blood cells. They can last for over a year. Sperm? About three days. Brain cells? A lifetime! But keep in mind those brain cells can't be replaced if they die early. Something to think about the next time you spark a big fatty.

I'll end with skin. Your skin accounts for about fifteen percent of your body weight and is composed of over 1.6 trillion cells. Even more amazing—you'll lose about forty thousand of those cells *each hour!* That's over a million each day! As new skin cells push their way to the surface, the ones on top die and eventually fall off (if you want to impress your friends, that dead layer is called the stratum corneum). The journey from the depths of the dermis to death on the surface takes about a month before those little skin cells flake off and flutter to the ground. In fact, take a look around your house or office. If there's dust, there's skin. Most of the dust that surrounds us is composed of dead skin cells, and you'll shed about eight pounds of them each year!

Our bodies produce a lot of waste, but it's to be expected. Each of us is composed of cells, tissues, organs, and systems, working in concert to keep us alive. And with all that productivity comes the discarded excess, the by-product of these amazing machines. Think about that the next time you take out the garbage.

THE SEEDS OF SYPHILIS
February 14

Syphilis. Few words strike such fear in the hearts (and other parts) of humans. Cancer, perhaps. Leprosy, for sure. But few diseases have such an intriguing history.

My introduction to syphilis began years ago. When I was a firefighter-paramedic, infectious disease (in various forms) was rampant among the transient and drug-addicted clientele of Orlando's west side. My tenure as a medic at a level one trauma center also gave me intimate insight into the syphilitic scourge. Ironically, retirement from the fire department brought little reprieve, for as a bioarchaeologist, I've come to know syphilis from a whole new perspective. Let me explain.

Syphilis is caused by *Treponema pallidum*, a little corkscrew bacterium that packs a mean punch. The common mode of transmission is via the mucous membranes of an infected person. The infection produces sores, which can also serve as portals for transmission. And don't think the sores are the worst part, for they serve as the jumping-off point of an infection trifecta.

Stage one (primary) syphilis commences with the appearance of these sores, which usually crop up at the point of infection. Keep in mind these sores can be hidden within the vagina or tucked discreetly in the anus, so just because the coast is clear, it doesn't mean your partner is syphilis-free (*Condoms, Condoms, Condoms!!*). The bacterium is happy to hide out in the bloodstream and then appear unannounced.

Stage two, or secondary, syphilis is marked by skin rashes accompanied by "mucous membrane lesions" (think mouth, vagina, and anus). The rash can adorn various body parts, from the palms of your hands to the soles of your feet, and ranges from severe to barely noticeable. Flu-like symptoms sometimes accompany the rash. Surprisingly, stages one and two will eventually clear up, even without treatment. But without treatment, the disease can progress. Stage three can be just around the corner or lie dormant for years.

Stage three (also called tertiary) syphilis marks the point where the disease affects internal organs. Muscles may grow weak and shaky, blindness can set in, and the mind starts to go. As it progresses, organs are eroded. The heart, brain, nerves, and liver fail, but it's the changes in the skeleton that excite bioarchaeologists, for these bony changes allow us to trace the disease back through time.

Syphilis, like any STD, initiates the blame game: no one wants to take credit for the origin of this nasty infection. This explains the plethora of names folks have conjured. The most common was the "French Disease." Of course, the French weren't too fond of this term, so they referred to it as the "Neapolitan Disease," which I'm sure infuriated the Italians. The Russians referred to it as the "Polish Disease," and so it went, on and on. . .

Prior to the advent of modern medicine, differential diagnoses (distinguishing one disease from another) were haphazard, at best. Historic descriptions of syphilitics predating the 1500s could be syphilis, could be leprosy. Who knows?? The diseases produce similar symptoms, and victims were traditionally banished from all good society, so it's difficult to tease apart their histories. To make matters worse, the *Treponemal* bacterium also causes other types of infections (bejel and yaws), which are spread by means other than sex. Talk about a contagion conundrum.

According to historic documents, syphilis spread rampantly throughout Europe during the 1500s—following closely on the heels of Christopher Columbus' voyages to the New World. So perhaps it wasn't the French, the Italians, or the Polish. Perhaps it didn't originate in Europe at all! There was only one way to know for sure: *Bioarchaeology to the rescue!!*

Syphilis leaves telltale lesions on the skeleton. It starts with small erosive wounds on the skull ("caries sicca," in bioarc lingo), which can then spread to the nasal area. The disease eats away at the bones of the face, destroying the nose and upper palate. It also affects the lower leg bones, or tibia, causing the deposition of extra bone that gives the tibia a curved, sword-like appearance, known as "saber shin." It's the combination of these lesions that indicates tertiary syphilis, enabling bioarchaeologists to differentiate it from other diseases such as leprosy, that other infamous bone destroyer.

By identifying syphilis on skeletons in archaeological contexts and radiocarbon dating these remains, we can trace its origin and diffusion across time and space. And after decades of debate, the current evidence indicates the disease was present in the Americas predating Columbus' entry into the New World. So far, no European skeletons predating the 1500s have been proven definitively to exhibit syphilis. However, there are bioarchaeologists still combing ancient remains in Europe to confirm these findings.

According to the most recent molecular analyses (DNA), syphilis mutated from bejel or yaws during its passage from Africa, into Asia, and finally into the Americas and was present in the New World by the time Columbus landed. So, Columbus not only brought tomatoes, potatoes, and corn back to Europe; but his randy sailors may have served as the first trans-Atlantic syphilis-smugglers.

The search for syphilis is a great example of the cooperative efforts of science. By using historic documents, skeletal evidence, and confirmation via molecular analyses, the origins of this hideous disease have been traced and the debate put to rest. At least for now.

A SYPHILITIC SEQUEL
February 21

Last week, we discussed the fascinating history of syphilis. If you didn't catch it (no pun intended), I advise you to give *The Seeds of Syphilis* a quick read.

We'll pick up where we left off: It's the dawning of the 16th century and syphilis is rearing its ugly head across Europe. The earliest reports of syphilis date back to 1495. The French army had marched into Italy, laying claim to Naples and, once the fighting ended, the soldiers settled in to celebrate their victory, treating themselves to the bevy of prostitutes who accompanied their expeditions. In no time, the soldiers were debilitated by fever, rash, and muscle pain. Over the next few months, the disease progressed, rendering them unable and unwilling to fight. Many were sent back to France, where they subsequently infected their loved ones.

By 1500, the "great pox" was spreading like wildfire. It accompanied da Gama's voyages to Calcutta in 1498 (the perfect side dish to chingri) and within twenty-five years, folks in Africa, Japan, and China found themselves grappling with genital sores. Even the islands of the Pacific fell victim, and for centuries, syphilis plagued much of the known world. So, let's take a look at some of the ways society dealt with this dreaded disease.

Following closely on the heels of the great pox, Jacques de Bethencourt coined the term "venereal disease" in 1527 to describe these "maladies of Venus," or as we now refer to them, STDs (sexually transmitted diseases). The word "syphilis" first appears in a 1530 poem. Girolamo Fracastoro (a true romantic) waxed on about the wretched symptoms of the disease, as well as some of the early attempts at treatment. These included an herb known as "holy wood," along with a strong dollop of mercury, which would persist as the primary syphilitic remedy well into the 20th century.

Mercury treatments came in several forms. Pills, ointments, and even luxurious mercuric steam baths were used to control the malady. The problem with mercury is it tends to be a bit poisonous. The side effects were sometimes as bad as the syphilis itself. Mercury caused its own fleet of ulcers, which typically cropped up in the mouth and throat; the patient's teeth tended to fall out; long-term use could lead to nerve damage; and worst-case scenario, the patient died of mercury poisoning. Not a great prognosis. It also gave rise to a catchy little saying:

"A night with Venus and a lifetime with mercury"

For a while, physicians toyed with inoculations. This process of "syphilization" was usually conducted on prostitutes, who were repeatedly dosed with "syphilis matter" (which I envision as the pustulating ooze from sores) in hopes of preventing the disease from advancing to its next stage. I wonder what the prostitutes had to say about this. . .

It wasn't until 1905 that the bacterium responsible for syphilis was identified. The Wasserman blood test (I'm assuming named after some guy named Wasserman) was developed a year later. Once a diagnostic test was available, clinicians no longer had to depend on visible symptoms. This helped reduce transmission, since many carriers unknowingly spread the disease.

As if treatment with mercury wasn't bad enough, the next remedy added arsenic to the mix. In 1910, the gifted scientist Paul Ehrlich developed Salvarsan, the first drug that actually attacked the bacterium. He tested his potent cocktail on syphilitic rabbits and knew he had struck gold when the bunnies survived! Salvarsan served as the primary treatment for syphilis. That is until our hero, Alexander Fleming, discovered penicillin in 1928.

And no syphilitic history would be complete without mentioning that most tragic of human experimentations, the Tuskegee Experiment. Begun in 1932 in Macon County, Alabama, the experiment was aimed at reducing syphilis among blacks. It was led by the Public Health Service in conjunction with the Tuskegee Institute and involved six hundred African American males—399 who had syphilis, 201 who did not—who were told they were being treated for "bad blood." The men were given free medical exams and meals, but those with syphilis were never actually treated, even after penicillin was widely available. The study lasted for forty years. Along with the exams, the men were given complimentary burial insurance which, sadly, many put to good use. The last participant died on January 16th, 2004.

Few diseases have impacted society the way syphilis has, but it has also been part of some of the most groundbreaking moments in medical history. Historic records, skeletal analyses, and molecular detective work have enabled us to track the disease back in time; improvements in diagnostics have streamlined the treatment process; and the use of drugs and education have drastically reduced its incidence around the world. So, let's hear it for syphilis! This little bug has really taught us a lot.

I'll leave you with the beautiful prose of Fracastoro:

"A shepherd once (distrust not ancient fame)
Posses' these downs, and Syphilus his name."
"He first wore Buboes dreadful to the sight.
First felt strange pains, and sleepless passed the night.
From him the malady received its name.
The neighbouring shepherds catch'd the spreading Flame"

BIRDBRAINS
February 28

Although I've spent the majority of my adult life swearing off pets, in December I broke down and officially became a pet owner. An avid birder for the past few years, I decided birds would be a nice addition to the house. I love watching them and listening to their calls, and they're pretty light on maintenance, so I took the plunge and bought a pair of lovebirds.

They're beautiful. I don't know their sex (DNA required), so I gave them boy names. Tuukka and André make a lovely pair (we are big supporters of gay marriage). They are named for two of my favorite NHL goalies: Tuukka Rask (Boston Bruins) and Mark-André Fleury (Pittsburgh Penguins) and I can tell you, for a pair of lovebirds, they certainly fight like hockey players.

I've spent considerable time training them over the past few months and they're finally to the point where they no longer remove a hunk of my skin when I pick them up, which is a big improvement. They've learned how to return to their cage on my signal, which they accomplish by running up a ladder (hilarious), and they are now content to sit on my shoulder and watch TV (they seem to enjoy the NHL).

They are very smart. Lovebirds are basically miniature parrots, which are known for their intelligence. But it amazes me that a brain that small can function so efficiently, especially since they lack the complexity of human brains. So, I did a little research to get a feel for what makes up a bird brain.

Bird brains contain many of the same structures as ours. As in human brains, their medulla controls basic bodily functions: heart rate, blood pressure, and respirations. Like us, they have an optic lobe, but theirs is more developed than other vertebrates since they rely strongly on vision. Their cerebellum, like ours, controls coordination and balance, but again, tends to be large, since it must coordinate the muscles involved in flight. And their cerebrum contains lobes, which tend to be larger in parrots and crows, compared to other birds. These lobes enable parrots and crows to be very adept with their beaks, which they use to manipulate objects. Parrots also possess agile tongues; a skill my lovebirds enjoy showing off by spitting shells out of their cage (which I'm pretty sure they do just to annoy me).

Tuukka and André have distinct personalities. Tuukka tends to be more aggressive and has been harder to tame than André. Tuukka entertains himself by shooting me the evil eye, while André is happy to hang out on my shoulder and nibble on my earring. Aggression in humans is controlled by the amygdala, which causes it, and the hypothalamus, which regulates it via receptors that interact with the neurotransmitters serotonin and vasopressin. Bird brains contain both structures, so I wonder if it's the same for them. If so, I'll have to see about getting Tuukka some serotonin supplements.

It was once believed that the three capacities separating humans from other animals were bipedalism, language, and tool use. Although we are the only obligatory bipedal primates, language and tool use present a vaguer division.

Studies among chimps and gorillas have shown they have the capacity for language, although at a more rudimentary level, and many animals communicate via distinct calls, which some consider a form of language. The copycat speech of parrots used to be considered simple mimicry, but new studies have shown that parrots can actually construct meaningful statements, count, and even understand the concept of zero. Alex, a famous parrot who was the subject of experiments in avian intelligence, shocked his assessors when he answered "none" when asked how many blue keys were among a group of green and red. He was right!

As for tool use, it is now well documented among the animal kingdom and it just so happens some of these animals are birds.

The woodpecker finch, found among the rocky landscapes of the Galapagos Islands, uses twigs to extract insects from the insides of trees. But he doesn't just use any old twig. He chooses an appropriate stick and trims it to the proper size before using it as a miniature pry bar to dig for bugs hidden within the bark. Crows show remarkable talent when it comes to constructing tools. They not only make tools but can preplan and problem solve—which is more than some humans can accomplish. And I've watched my birds use their beaks to hoist objects out of their way, which makes them not only smart little suckers but strong, to boot!

And I'm convinced my lovebirds dream. Every now and then, in the still of the night, one of the birds will utter a blood-curdling "Chirp!!" before settling back to sleep. What is he seeing in his little bird nightmare? A pouncing cat? My hand reaching into his cage? What frightening images do their little brains conjure?

Birds and primates parted evolutionary ways over 280 million years ago and as we evolved along our separate paths, so too did our brains. As humans, we may be intelligent, but we certainly don't own it. Pigeons can memorize over seven hundred different visual patterns. Scrub jays exhibit episodic memory. Owls learn the many night sounds that lead them to meals. And African grey parrots can understand numerical concepts—a skill once thought to be uniquely human.

So, the next time someone refers to you as a "birdbrain," take it as a compliment. Birds are pretty darn smart.

THE SIDESHOW
March 7

This week, I started a new job. For those of you who have been reading the blog for a while, you'll recall back in May's *Losin' It* that I lamented the inevitable demise of my position. Well, it ended last month, my new position started Tuesday, and all is well in the world again.

Jobs are important. Think how much time we spend at our jobs, how they can define who we are as individuals. When I left the fire department, I went through a serious identity crisis. For thirteen years, I proudly wore the title "firefighter-paramedic." I was a "hero," a "lifesaver," a self-proclaimed badass. When I left to enter grad school, I was suddenly a nobody (the title "grad student" won't even get you a cup of coffee).

But when I graduated six years later and marched from campus with my shiny new PhD, I was somebody again, with a brand-new identity. I admit, going from firefighting to archaeology was a strange transition and, although my careers were at odds with each other, they made for some incredible experiences.

So, in recognition of odd jobs, I thought we'd take a quick look at the disturbing history of human exhibitionists, otherwise known as the "circus freaks."

That common term was a cruel label given to those who relied on physical deformities or biological oddities to eke out a meager living as a circus sideshow. During their heyday in the 1800s, these unfortunate individuals were readily labeled "freaks," since the medical community lacked the sophistication for accurate diagnoses.

For example, Lionel, The Lion Faced Boy was covered with six-inch-long hair over most of his body. Born in Poland in the late 1800s, Lionel suffered from hypertrichosis, a rare genetic disease that causes excessive hair growth. Today, this condition is treated through medications that inhibit hair growth or through manual hair removal, but no such treatments were available to poor Lionel. He died of heart failure at the age of forty-one.

Conjoined twins were always a big draw and Chang and Eng Bunker were two of the most famous. Born in 1811 in Thailand (then known as Siam, thus the name "Siamese Twins"), the brothers were omphalapagus twins—joined at the abdomen—and shared a single liver. Had the brothers been born in the modern era, they most likely could have been separated. Separation depends on the type of twinning involved and omphalapagus twins, which account for about thirty-three percent of conjoined twins, have a high success rate when the heart is not involved. Fortunately, the brothers' condition didn't seem to slow them down. They married a set of sisters and ended up having twenty-one children between them. Imagine that!

Another condition that resulted in deformities deemed worthy of exhibition was acromegaly. The condition causes excessive growth of various body parts, typically the hands, feet, and jaw, as a result of overproduction of growth hormones (usually from a tumor of the pituitary gland, which controls such hormones). But it also causes thickening of the skin, excessive height, and an enlargement of the bones of the face. The deformities can be frightening. Mary Ann Webster suffered from this horrible condition. Born in London in 1874, Mary, like many sufferers, developed the disease as an adult (the condition is called "gigantism" when it strikes children). As the disease progressed and her face became more distorted, she was cruelly billed as "World's Ugliest Woman" and put on display.

Of course, the most famous sideshow star was Joseph Merrick. Dubbed the Elephant Man due to the leathery grayness of his skin and its rough, mottled texture, along with the gross deformity of his face and body, Joseph lived a tortured life. Shunned as a child and suffering from what has now been diagnosed as Proteus syndrome, he went into the sideshow business after several failed attempts in sales. But as the shows fell out of fashion in England, his handlers sent him on tour throughout Europe, where he ended up beaten and robbed. After making his way back to London, he was taken in at a local hospital and, through the generosity of the medical staff and an outpouring of public support, was granted a permanent room where he lived out his short life, dying from asphyxiation at the age of twenty-seven.

Humans can be cruel. Even under the best circumstances, our differences can evoke ridicule and abuse. So, the next time you have a bad hair day or bemoan the fact you've gained a bit of weight, grab an ounce of perspective. Think for a moment what it must have been like for those who made their living off the public's painful scrutiny and for those forced to carry on this horrific tradition today. May they find peace.

EATING OUR OWN
March 14

In the early 1900s, anthropologists studying aboriginal populations on the Pacific Island of New Guinea documented a strange disorder among the folks known as the Fore. The symptoms were most common among Fore females, and began with tremors, slurred speech, and difficulty walking before advancing to loss of muscle control, dementia, and death. The scientists were stumped. Genetic? Probably not, since such a lethal anomaly would have been weeded out of the small gene pool.

They finally traced the disease to a virus, but it was the mode of transmission that was most disturbing. The virus was contracted by members of the tribe who were eating the infected flesh of their dead.

The Fore participated in a unique mortuary ritual. When a family member died, the kin would meticulously dismember the corpse, remove the entrails, and scrape out the brains. This gruesome task was carried out by females, who refused to let all that meat go to waste—especially the brains. When the brains happened to belong to an infected individual, the virus was passed along to the hungry relatives. Thus, the mystery was solved and cannibalism, the culprit.

Cannibalism, known in nerd-speak as "anthrophagy," is nothing new. There is even evidence, in the form of cut marks on bone, that Neanderthals may have been consuming other Neanderthals. Humans certainly have a long tradition of eating their own, and there is a range of reasons for snacking on *Sapiens*.

We'll start with the traditional forms of cannibalism. Although claims of cannibalism are probably overstated, there are documented accounts. But even cannibals have preferences.

Some will only eat outsiders—what we term, *exocannibalism*. For example, following a battle, some groups will consume their fallen foes as a statement of power. It was reported in 2003 that Congolese rebels supposedly ate the bodies of pygmies taken in battle. Some cannibals do it simply for sport. The Mianmin of New Guinea would hunt down neighboring tribes when they craved an exotic treat.

Endocannibalists are those who restrict their consumption to members of their own group. It's traditionally tied to spiritual beliefs; a way of holding on to the dead or acquiring aspects of their personality. Perhaps they view it as a form of "comfort food." Like Americans and their fried chicken.

Some forms of cannibalism are part of a broader deviancy. We've already discussed Jeffrey Dahmer's exploits in October's *Dead and Lovin' It*, but Jeff isn't alone in his quest for flesh. Here's a case from my home state of Florida. In May of 2012, police shot and killed Rudy Eugene after he was found naked on the interstate munching on the face of an elderly homeless man. To this day, no one knows why Mr. Eugene suddenly turned cannibal (although I believe Miami brings out the "weird" in all of us).

In the 1600s, cannibalism was part of the early medical landscape. It was believed the pulverized flesh of Egyptian mummies contained curative properties, thus medicinal cannibalism became a widespread practice across Europe, persisting into the 1900s. But medicinal cannibalism predates by a long shot the "mummies as medicine" approach. Galen, one of the founding fathers of medicine, prescribed human blood to treat a range of disorders. Of course, he also believed blood flowed through two separate systems in the body and venous blood was produced in the liver, but even geniuses get it wrong sometimes.

But back to cannibalism. There is much debate among anthropologists about the accuracy of many accounts of cannibalism. During the era of colonialism, the accusation of cannibalism was a means of categorizing a group as subhuman, as monsters. It was much easier to justify enslavement and genocide when those you were capturing or killing were lowly "eaters of the flesh."

But cannibalism has also been undertaken out of sheer desperation; in some cases, fairly recently, where people have been forced to choose between cannibalism and starvation. This is known as survival cannibalism, and those who partake are compelled by that most basic ultimatum: eat or die.

New evidence points to cannibalism among the Jamestown settlers who came to Virginia in the early 1600s. Bioarchaeologists identified cut marks and signs of dismemberment on the remains of a teenage girl. Out of the original three hundred settlers, only sixty survived what became known as the "starving time" —the intense winter of 1609.

A more recent incident occurred in 1972 when a Uruguayan rugby team flying to Santiago, Chile, crashed high amidst the Andes Mountains. Of the original forty-five passengers, some of whom were killed in the crash, sixteen managed to survive for over two months on the barren mountain by eating the flesh of the dead.

And who can blame them? Yes, there will be those who claim they would rather die than eat a fellow human, but no one can truly say what they would do in such a situation. As far as I'm concerned, meat is meat and survival a heck of a better alternative than death.

Besides, consider some of the fast-food garbage we enthusiastically consume. A human's gotta taste better than that.

A VIEW FROM THE WOMB
March 21

Close your eyes and imagine a list of all the places you've ever lived in your lifetime. Organize the list however you choose: states, countries, continents. Perhaps your list is short, although it's a rare individual who is born and raised in one place these days. I grew up a military brat, moving about every two years. By the time I was a teen, I was acclimated to the life of a transient.

Although our backgrounds may vary, there's one place all of us have lived; one location that is central to each of our existences. We have all inhabited that magical, muscular pouch, the womb.

Known in technical circles as the uterus, this hollow, expandable sac resides just north of the vagina and serves as the site of incubation for each and every one of us. The inner lining (endometrium) is a thick layer of tissues that grows ever thicker as it prepares for pregnancy; the optimum thickness of a healthy uterus is eight millimeters, in case you were wondering. This blood-rich lining will nourish the fetus on its path to personhood while providing a cozy-comfy abode in which to grow.

But what's it like in there? When you spot a pregnant woman, do you ever wonder what life is like for the fetus? What sounds and sensations assail them as they float within that somnolent sea of amniotic fluid? Let's spend a few minutes inside their world.

As early as week seven, that little peanut is moving around. The mouth and tongue are almost developed and although the taste buds won't develop until week twenty, the fetus is sucking and swallowing about a liter of amniotic fluid per day. This will prepare them for real-life feeding later on. A newborn as young as three days old can differentiate between sweet, sour, and bitter tastes. It can also tell the difference between breast milk and formula and, I'm betting, would overwhelmingly choose breast over bottle.

By week eleven, the sensory nerve endings are in place and by the fourteenth week, innervation is complete, meaning the wiring that enables the tot to experience and react to its surroundings is up and running. By week sixteen, the fetus will usually begin to kick, which is known as quickening. (Is it me, or is does that word conjure up a scary image of a demon creature?)

By the twenty-fourth week, the ears are developed and functioning. I wonder what the fetus can hear—I suppose the muffled sounds of a subsurface world, like what a frog experiences as he cruises the depths of his pond. I'm sure there's a constant soundscape of bodily functions as the mother churns out the metabolic necessities for that demanding little tenant, and I bet by the ninth month, the fetus has become immune to the ever-present rumbles of flatulence. Pregnancy is a gassy business.

And speaking of gas... I wonder what a womb smells like. The nasal structures are in place by week eight, although that sense of smell will really kick in once the baby makes its debut. Smell is one of the most developed senses in a newborn and it enables the little tyke to recognize its mother and root out the nipple. Newborns can even sniff out danger, although their defenses are limited to a grunt and a squirm.

The visual system is one of the earliest to begin developing but takes the longest to complete. The eye begins forming as early as the third week of pregnancy (even before the lids can close), and the optic nerve, by week eight. By the sixth month, the visual cortex is innervated and by month seven, the little bugger is taking a look around. But the eyes aren't complete at the time of birth; vision will continue to improve, even as the newborn reaches his fourth month. And if the baby is born early, its eyes may still be sealed, like a puppy's.

It's a shame we can't remember the womb. What safer place could there be than tucked inside your mother, blissfully unaware of the harsh world that awaits your arrival; protected from the noise and stress of everyday life, all your needs met without having to lift a teeny, tiny finger.

I'll never experience growing a live person. I decided long ago to forgo parenthood, but I have been on the receiving end of pregnancy. During my tenure as a medic, I had the good fortune of delivering five healthy babies, so, although I didn't take part in their development, at least I ushered them safely forth. And that's about as close to pregnancy as I care to come.

NAVEL GAZING
March 28

Stop what you're doing. I want you to perform a quick experiment. Slide your hand under your shirt and stick your finger in your belly button (for those of you with "outies," please play along). I want you to contemplate that little crater and how, for nine months, it served as the life link to your mother, providing all your metabolic necessities. Let's face it—the belly button gets very little respect.

Last week's *View from the Womb* took us back to our fetal origins, so I thought it only appropriate to discuss how each of us was nourished and sustained during the sojourn in our mother's belly. When you think about it, the belly button is an amazing little nugget of mammalian anatomy.

Humans are placental mammals, which means each of us develops within our mother's body and is delivered alive and kicking (hopefully) to the outside world. As opposed to being hatched from an egg or transferred to a pouch, kangaroo-style.

The belly button serves as the point of attachment for the umbilical cord which, together with the placenta and the amniotic sac, make up the life support system for the developing fetus. The cord itself contains three main vessels: two arteries that deliver deoxygenated blood from the fetus to the placenta, where it is oxygenated, and a vein that returns the oxygenated blood back to the fetus. No, you are not confused; the roles of the arteries and veins are reversed, just as in our pulmonary system.

The cord sprouts around the third week of pregnancy and can grow up to sixty centimeters (about twenty-three inches) long, allowing the fetus to perform its amniotic acrobatics during pregnancy. In the case of twins, although they share the placenta and may even occupy the same amniotic sac, each will have its own cord (the one instance where it's not nice to share).

The umbilical cord allows the exchange of nutrients, like amino acids, glucose, and oxygen, between the mother and fetus while also serving to eliminate waste, such as carbon dioxide. The placenta does the heavy lifting of fetal nourishment; the cord simply acts as a conduit.

When the baby is born, it is still attached to the placenta via the cord. The cord is cut as soon as it ceases pulsing (a few minutes after birth), and the baby is left with an ugly little stump, which will dry up and fall off about two weeks later. Your fate as an "innie" or an "outie" is determined by the amount of scar tissue that develops. Extra tissue means you'll join the ten percent of folks who sport an outie.

Blood from the umbilical cord is a precious commodity. It contains valuable stem cells, which doctors believe will eventually be used to regenerate tissues. If your liver fails, they may simply grow you a new one! Cells from the cord are also showing promise in treating burns. The cord is a gift that keeps on giving.

It turns out there's been quite a bit of navel science in recent decades and the research is providing some interesting button-info. For instance, they've identified over fourteen hundred different strains of bacteria nestled within the navel; something to think about the next time you're running your tongue over your partner's belly. And depending upon how deep the button delves, there may also be quite an accumulation of lint. The bigger and hairier the belly, the greater the lint accumulation, since it's the rubbing of belly hair against clothing that produces the fuzz. It pays to do a bit of housekeeping from time to time.

Duke University researchers have found a correlation between the height of the belly button and how fast one can run or swim. Turns out the higher the button, the faster the athlete. Who'd a thunk it?

The size and shape of the navel is also associated with sexiness. Supposedly, a shallow button with a slight hooding is considered more appealing than other manifestations. Outies are out, as are those that are too deep. Mine's a bottomless pit, so I guess I'm relegated to a life of social isolation.

And if you are pregnant, you can morph from an innie to an outie. But have no fear: when your body returns to normal, the button usually follows suit.

Each and every aspect of our bodies represents who we are and where we came from—from our evolutionary past to our reproductive present. Our belly buttons serve as links to our fetal past, tangible evidence of the connectivity between mother and child.

So, the next time you squabble with your mom, take a deep breath, and fondle your button. Always remember who got you here.

HAPPY BIRTHDAY TO ME
April 4

Happy Birthday, Body Blog!! You are one year old this week. If you were a human, you would be graduating to the realm of toddler, you'd have tripled your birth weight, and you'd be learning how to grasp objects. But since you are just a blog, we'll focus on the lesser achievements you've made over the past twelve months.

A year ago, I didn't even know what a blog was. I assumed they were merely outlets for people to spew their unadulterated opinions across the Web; kinda' like Fox News. But over a cup of coffee with a genealogist friend of mine, I learned that blogs could be educational outlets for writers, so I delved right in. I learned all I could about design and content and then I started writing, and for every week since, I've tried to bring you interesting tidbits about these amazing vessels we inhabit.

It was slow going at first. I had recently set up a Facebook page to help market my books but didn't really understand the subtleties of social media. But once I caught on, so did the blog, and my readership has improved by leaps and bounds.

But why write about the body, you ask. Well, I've nursed a fascination with the human body my entire life, from its underlying framework of bones and muscles to the intricacies of our internal organs. I knew I would end up in some realm of the medical field. I had no idea it would be as a firefighter-paramedic, nor that I would transition later on to bioarchaeology. But you know what they say: life is a journey, and mine has taken many a strange detour.

Writing about the human body holds endless possibilities. The wonders and mysteries of the human form are limitless, and I felt my unique perspective as a paramedic, archaeologist, and anthropologist would provide numerous angles from which to analyze and contemplate.

And you, as readers, guide my subject matter. I see what topics stimulate, which posts get shared the most and the farthest, and I take that into consideration when planning my subjects. But I don't let these things dictate. If I did, I'd end up writing about nothing but the lady parts, since they seem to be THE hot topic among Google searches.

Birthdays are a time for reflection. When my birthday rolls around (which they seem to do with ever-increasing frequency), I can't help but stop and reflect on the year past and the year ahead. A compulsive planner, I use my birthday to evaluate current life strategies—to make adjustments to those that aren't producing and set new objectives for the coming year. Yes, I approach life with a robotic precision (I would have made an awesome drill sergeant), but it's this fanatical foresight that has enabled me to achieve more than I ever imagined I could.

Is it genetic? Part of it, I'm sure, is written in my DNA, for this compulsiveness spills over into other realms—I'm a raging germophobe and excruciatingly neat. But part of it has developed through life experience. I wasn't always so driven. Everything changed with the death of my mother.

She was only fifty-two when she died of cancer. I was twenty-three, with my whole life ahead of me, but as I got older and realized how quickly the years fly by and that, like her, I could go at an early age, I entered into a subconscious game of beat-the-clock. Suddenly, I was an adult, racing towards thirty, and although I had an awesome position on one of the best fire departments in the state, it wasn't enough. I stayed in school throughout my career with Orlando Fire Department, knocking out degrees one by one in a race to achieve. And when I hit my ten-year mark and was vested in my pension, I realized if I was ever going to try out another profession, I had better get to it. I wasn't getting any younger.

On a whim, I applied and was accepted into grad school at Florida State, which was fortunate, for when my career at OFD took a sudden turn, I was prepared. I retired, packed up my life, and headed to FSU, embarking on a whole new career, a whole new life.

I've been really fortunate. All of my compulsive planning, my demented discipline, and tireless work ethic have paid off in big ways. I've achieved more than I ever dreamed I would when I was a young firefighter riding backwards on an engine.

So, as I reflect over the past year of writing The Body Blog, the moral of my story is this: set goals and work hard. If a dorky firefighter like me can go on to achieve a PhD, anything is possible. I think my mom would be proud.

OUR VERSATILE VAULTS
April 11

When I was six years old, my mother gave birth to her fourth child. Expecting another girl, we waited in anticipation for the emergence of our new sister. Alas, it was not to be. Out popped a strange, crinkled raisin with even stranger crinkled raisins between his legs. Our baby brother, Andy, was born.

Not only did he have a strange set of genitalia and a disgusting little stump of a belly button, but he also sported an intriguing soft spot at the top of his head. Since I was just a tyke with little understanding of cranial morphology, I couldn't quite grasp that the flexible area at the top of his head, which bulged whenever he screamed and strained, was a necessary component of skull anatomy.

The soft spot, or fontanel, is the area at the top of the newborn's head where the cranial bones have yet to join. There are actually six primary fontanels that reside between the bones of the skull; the largest is on top and usually fuses at around eighteen months.

Let's pretend, for a moment, you were born without them. Without the fontanels, you'd probably still be dangling from your mother's vagina, since there's no way that giant fetal head is going to make it out without a flexible skull. Through the magic of evolution, nature has prepared us bigheaded humans for the tight squeeze that is the birth canal by postponing the joining of the cranial sutures until *after* we've cleared the lady parts.

The cranial vault is not a singular bone. It is composed of a number of bones that are joined via the forementioned sutures. The sutures fuse over time, becoming completely obliterated if you're lucky enough to reach old age.

So now that we understand the basic mechanics of the vault, let's explore some of the clever ways we humans have manipulated our malleable melons. Welcome to the magical, mystical world of cranial deformation.

Beauty comes in a variety of forms and is highly dependent upon culture—what one group finds attractive may repulse an outsider (nipple rings make me want to hurl). When it comes to adornment or decoration, the human body provides a veritable canvas upon which to express ourselves, and humans have been modifying and mutilating themselves for tens of thousands of years.

Because a newborn's head is so flexible, it doesn't take much to bend and shape it into a variety of strange yet eye-catching forms. For the ultimate in head-shape handiwork, let's travel back a few thousand years to Central America and take a quick peek at the Maya, for they were rather deft at deformation. (Yes, I can spew these catchy phrases all damn day!)

The Maya was a sophisticated culture that thrived in Mesoamerica for thousands of years before experiencing a rapid decline, although their ancestors can still be found around the Yucatan. They were a learned society, with a written language and math, and they also built some pretty impressive pyramids.

And when they weren't calculating their celestial calendars or playing their oh-so-serious ball games (the losers got death), they were sculpting their newborns' heads into a variety of beautiful shapes and sizes.

According to the Spanish (who, upon seeing such oddly shaped heads, took a break from pillaging to ask a few questions), the reason the Maya constructed such elaborate shapes was to appear more handsome before their gods. They also made it easier to carry stuff. The broad, flattened variety made an ideal shelf for shuttling baskets from market to home. Very practical folks, the Maya.

Deformation was achieved via two basic techniques, which could be altered for more variation. The head could be bound with tight wrappings, and the location of the binding would determine the direction and shape of the deformation. The other technique employed specially designed cradles, which squeezed the head in a desired direction. Either method could not have made for a happy baby. Imagine your young'un strapped to a board for hours on end. The crying must have been relentless. No wonder the Maya were big on human sacrifice.

Deformed heads were considered beautiful by the Maya, an indication of rank among a sea of common noggins. And many of these skulls have been recovered among the archaeological ruins of this most impressive civilization. But the Maya weren't the only folks to dabble in deformation. Evidence of intentionally modified heads has been found on nearly every continent going back tens of thousands of years.

Sadly, the custom has fallen out of favor, for what I would have given to strap my baby brother down, sit back with a bowl of popcorn, and watch his little head elongate into cone-headed perfection.

BRAINY BONDAGE
March 18

Last week, we discussed the fantastical flexibility of the newborn's skull and how cultures around the world have taken advantage of its malleability by intentionally modifying it into strange and mesmerizing shapes.

This week, however, we are exploring the opposite end of the spectrum: what happens when your brain is locked inside a box that cannot yield. Let me tell you up front—it ain't good.

By the time a child is around three years of age, the fontanels, or soft spots that reside between the plates of the skull, have vanished as the sutures joining the plates fused. Those sutures will continue to meld throughout your lifetime and, if you live long enough, may completely disappear (what we in the skeletal biz call "obliterate").

But having a network of fused plates encasing that most essential organ, the brain, is a double-edged sword. Let's discuss the pros and cons of this brainy box.

The skull offers protection to the delicate tissues that compose the brain. Without the skull, you'd most likely suffer brain damage before lunchtime, for without the protective cover of the cranium, all it would take is a subtle knock on the noggin to cause permanent damage. There's a reason all vertebrates sport a helmet at the northernmost region of their spinal apparatus. Whether you're a tiger or a toad, an elephant or an anteater, protection is a must, and the cranium does a pretty fair job of buffering our brains.

But even with a skull, damage happens. A trip and a fall or a crash on your bike is all it takes to cause a head injury and one of the scariest injuries you can suffer is closed head trauma.

Closed head trauma is just as it sounds: the brain is damaged even though the skull is intact. The reason these types of injuries are so serious is that the brain, like any other soft tissue in the body, tends to bleed and swell when damaged. And whether the brain bleeds, swells, or both, there's no place for any of it to go. It is trapped within an unyielding bony container and without immediate intervention, the pressure intensifies, neurochemicals are released, causing further damage, and basic functions (like breathing and heart rate) are stifled as the brainstem is choked off.

So, what's a neurologist to do?? There are drugs to reduce the swelling which are usually the first line of treatment, but if the swelling persists, they must take further action. Rev up the bone saw; it's time to make a window.

Craniotomy is just as daunting as it sounds. A flap of bone is removed, which allows a bit of breathing room for the swollen brain. The piece of bone is usually popped in the freezer so that, when the brain stops swelling and returns to normal, the bony patch can be replaced and the patient restored (one hopes) to normal function, or at least something close.

The most amazing aspect of craniotomy is that folks have been performing it long before modern medicine came on the scene. It turns out we've been cutting into each other's skulls for thousands of years.

"Trepanning" is from the Greek, meaning to "bore" or "auger," which is most appropriate, since practitioners over seven thousand years ago were drilling into the heads of their friends and neighbors. Also known as "trephination," the procedure was originally accomplished using flint scrapers, knives and even shell; that is until the advent of metallurgy. Obsidian blades were also effective, as were bow drills. We can sometimes determine the tool used by the evidence left on the skull. A round hole with beveled edges? Probably scraped. Square hole with straight lines? Probably a blade. A beautiful little circle? Bet on a drill. Not only can we decipher the tool used, but by noting healing (remodeling) to the periphery of the wounds, we can also determine if the individual survived the procedure, which, surprisingly, many did.

The ancients may have been trepanning for some of the same reasons we perform craniotomies, since some of the skulls show evidence of traumatic injury. Was trepanation done solely for medical reasons? Probably not. There may have been a ceremonial component to it, perhaps to release bad spirits or correct psychotic behavior. Perhaps it was to obtain bone amulets, since it was performed on the dead as well as the living, and nothing says "fashion" like sporting a bit of cranial bling.

Whatever the reason, we know that people around the world for thousands of years were dabbling in neurosurgery long before there were neurosurgeons. Just as early explorers wondered what lay beyond the horizon, the human body also afforded a landscape worth exploring. And what better place to start than a hole in the head?

THE MOON'S PULL
April 25

Last week, nature graced us with a lunar eclipse. Having never experienced one, I was all atingle as I anticipated this exciting cosmic event. I set my internal alarm clock for 3:45 a.m., when the eclipse would be at its peak, woke up right on the button, and headed outside. I stood in the middle of my dark street, bleary-eyed in my pajamas, and stared up at a beautiful dusky moon.

It was a perfect night for it. The sky was clear and cool, rimmed in faint stars that paled in comparison to that full, glowing satellite. Its face had taken on a reddish hue, as if the entire surface had been dusted with cinnamon. It was a remarkable sight that has stuck with me all week.

It got me thinking about the mythical effects the moon is believed to have on the human body. We've all heard the tales. A full moon purportedly ushers in a range of bodily reactions, from erratic behavior to seizures. It's even rumored to manipulate menstrual cycles! As a new paramedic, I remember the warnings from seasoned personnel about the nutjobs that would emerge when the moon was full (although it seemed my fellow firefighters were far more afflicted than any of our patients). So, I thought we'd have some fun sampling these old wives' tales and, hopefully, lay some of them to rest.

The lunar myths are tied to its gravitational pull. Folks believe that since the moon regulates the tides on earth, it should have similar effects on the human body since, like the oceans, we are composed mostly of water. The only problem with this theory is, the last time I checked, the oceans were a lot bigger. And besides, the tides go in and out despite the phase of the moon. Yes, a full moon can cause subtle surges in tides, but the same effect happens with a new moon. It's not so much the phase of the moon as it is the lining up of the sun, earth, and moon, which occurs during both the full and new moons. So, rest easy. . . your bloating should not intensify with the waxing moon.

As for the moon's effect on seizures, epilepsy is a disorder characterized by unpredictable seizures that affects people of all ages. Caused by erratic electrical events within the brain, rumor has it these events intensify under a full moon. But epileptics—have no fear. When scientists reviewed the frequency of seizures over a three-year period, they found no increase when the moon was full. Quite the contrary: more seizures occurred during the last quarter, so if you're going to stockpile your Dilantin, do it for when the moon wanes.

And what about menstrual manipulation? Apparently, someone (gotta be a man) came up with the idea that the full moon somehow influences menstrual cycles, I guess since they both occur about once a month. However, few women flow with as much regularity as the lunar cycle and, if the moon actually did influence menses, then in my mind it should cause a world-wide synchronicity resulting in global premenstrual syndrome, aka, Armageddon!! God help us!

Emergency room visits have been rumored to increase beneath a full moon, as have animal attacks. Could it be these events are related? Seems only natural that if the critters are biting, the ERs would be buzzing. But once again, science has dispelled this myth. There appears to be no increase in ER visits when the moon is full. As for critter bites, a British study back in 2001 found twice as many animal bites when the moon was full, but studies elsewhere fail to corroborate their findings. Perhaps it's just a "Brit" thing. Have you seen the size of the rats in London??

What about the psychological effects of the full moon? The poor moon is blamed for everything from depression, to suicide, to psychotic outbursts. In fact, the word "lunacy" comes from *Luna*, the Roman Goddess of the Moon, and the condition was believed to be linked to the moon's phases.

We can thank Pliny the Elder for that one. Although Pliny was a deft naturalist, he missed the mark when he claimed that, since the full moon caused an unusually heavy morning dew, it must have the same effect on the brain. Pliny believed it was the "unnaturally moist" brain that led to erratic behavior. I'm no Einstein, but it seems logical that a moist brain beats a dry brain, any day. Alas, old Pliny never got to see his theory falsified; he was among the tens of thousands killed when Mount Vesuvius blew.

The only bodily effects possibly attributable to a full moon are subtle changes in sleep patterns. Studies hint that people may stay up later on nights of the full moon, which can subsequently interfere with normal moods and temperaments. This might certainly have been the case among our distant ancestors, who were more in tune and reliant upon the light of the moon than us city dwellers in our climate-controlled apartments.

And no discussion of the full moon is complete without the myth-of-all-moon-myths: werewolves. According to the website, "Mythical Creatures Guide," the belief in the transformative properties of the full moon and its resultant "manwolf" is widespread throughout Europe and comes in a variety of tall tales. Some werewolves were the result of witchcraft. Some were branded at birth by being born with a caul over their faces. And some were simply randy corpses who returned from the dead for conjugal visits with their widows. However, just as our search for Bigfoot keeps coming up empty, so have our quests to locate a bona fide werewolf. But it sure is fun to imagine.

Picture this: The bright, full moon slowly rises in the east as the metamorphosis begins. The man's canines elongate, his fingers morph into claws, he starts to drool as dense hair sprouts from every inch of his body. . . Wait a minute, I think I dated this guy!

SELECTING FOR SEX
May 2

My first kiss took place beneath a lunchroom table. The object of my affection was Whit Winfrey, a beautiful, blonde boy with striking blue eyes and tan arms adorned with feather-soft, white hairs. I lured him under the table to show off to my girlfriends, who had the audacity to doubt my ability to snag a kiss amidst the glaring gaze of the lunchroom monitor. I succeeded, Whit was baffled, and I smugly returned to my sandwich as my girlfriends looked on in awe. Such is love among fifth graders.

It amazes me that I can remember such detail about Whit's beauty, especially since many of my boyfriends have faded into the dark recesses of my memory. Whit was the first in a long line of fair-haired boys (with a few brunettes sprinkled in for variety) and even today, I'm a sucker for the blondes. But why? Why do we prefer certain physical traits over others?

Close your eyes and think about what turns you on (please keep your hands where I can see them). Are you a breast man or is it long legs that float your boat? Ladies, do you like a hairy chest or prefer a baby-smooth landscape? We all have certain physical traits that appeal to us more than others. But what you may not have considered during your lustful meditations is the role attraction plays in evolution.

There are three primary elements of evolution. The first is mutation, commonly referred to as "descent with modification." Without subtle changes in the genetic code, the second element, natural selection, would have nothing to work on. A fixed and unchanging organism wouldn't last long in the dynamic environment of earth, and it's the chance mutations that happen to confer an advantage that enables them to take hold in a population.

The third element is sex, or more specifically, sexual selection. But sex is nothing new; it's been around for about a billion years.

Prior to the advent of sex, organisms reproduced asexually and many still do today: bacteria, protists, some plants and fungi, and even some animals. There are many ways to do it (or should I say, not do it): fission, budding, and spore formation, to name a few. There's also parthenogenesis, in which the embryo develops without fertilization. If you were an aphid, a water flea, a hammerhead shark, or a komodo dragon, you'd be able to fly solo when it came to procreating. Think of all you'd be missing. . .

The development of sex was a boon for evolution. It is believed the recombination that occurs when organisms reproduce sexually, and therefore exchange genetic material, may purge bad mutations from their DNA. Recombination also provides a buffer against harmful mutations, since you have a fifty-fifty chance of getting a "normal" copy of the gene from one of your parents (which is why inbreeding is not such a good idea). Sex may also help fight disease by promoting the evolution of new genetic defenses. So, to state the obvious: sex is good.

But a big part of sex comes down to choosing a mate. Here's where things get really interesting. How and why we choose whom we choose is based on what we find attractive in a mate. For the most part, it's all about "fitness." Fitness, in the biological sense, refers to an organism's ability to leave behind offspring. Fitter individuals, be they moose or mollusks, will leave behind more moose or mollusks than their less fit counterparts and, in many cases, it's the females who will do the choosing. As you can imagine, fitness has a broad range of expression, depending upon which species you belong to.

Take that moose, for instance. A female moose is wise to choose a male who's muscular, and therefore can outrun predators, and one who sports a large rack (bet you never thought you'd hear that in reference to a male). That large rack will enable him to out-compete rival males as they vie for females.

But here's the conundrum: the wild and whacky world of sexual selection appears in many ways to conflict with the basic premise of evolution. Natural selection is supposed to favor those who are better adapted to survive, yet many physical attributes that crop up as the result of sexual selection actually reduce chances of survival by encumbering their owner with extraneous or physiologically expensive accoutrements (the peacock's tail is the typical example).

A moose may develop an enormous rack, which gives him an edge in combat displays, but it's also pretty darn heavy, making it more difficult to outrun a predator. Same with that peacock. He may strut around and display his tail to lure a prospective hen, but try evading a coyote while dragging that feathery aphrodisiac. My money's on the coyote.

Natural selection has come to terms with this predicament. If an individual can sport such elaborate ornamentation (be it a giant rack or a monstrous set of tail feathers) and still survive and reproduce successfully, that individual passes the ultimate fitness test and will usually go on to leave more offspring.

As for humans, fitness is still defined by physical traits that represent mating potential, whether it be a beefy, masculine male or a voluptuous, fertile female. But we are also cultural beings, and today, culture dictates much of what we consider desirable in a mate. And among us westerners, seems all you need is a flashy car or enormous breasts and you're in like Flynn. Happy hunting!

181

SEX EVOLVES
May 9

About a billion years ago, two organisms floating within the primordial soup bumped into each other and had sex. Voilà! The revolution that would be sexual reproduction was born and organisms throughout the world have been coupling ever since. OK. . . it was a bit more complicated than that, so we'll break it down.

The world was a very different place when sex first arrived on the scene. The ozone was just forming, providing some protection against the sun's raging UV radiation; oxygen was up to ten percent, about half of what it is today; and the continents were clustered in a giant clump known as Rodinia. The planet was inhabited by single-celled organisms, primarily green and red algae. Needless to say, it was a pretty monotonous place.

But big things were happening. The eukaryotes, those cells sporting a nucleus, had evolved and multicellular critters were about to burst onstage. Fast forward half a billion years to the dawning of the Paleozoic and suddenly, life as we know it (or at least life as the trilobites would know it) was in full swing. And what role did sex play in all of this? *A big one!!*

Scientists are still puzzling out how organisms transitioned from asexual to sexual reproduction, but they take hints from the natural world. Some bacteria are able to exchange genetic information on a rudimentary level, and many critters, such as yeast, are able to switch back and forth between the two methods as conditions warrant. So, the basic evolutionary mechanics of sex are slowly being teased apart.

Once it kicked in, sexual reproduction fueled natural selection, as it still does today. Instead of organisms making carbon copies of themselves and relying on the occasional mutation to provide an edge among their competition, sex added valuable variation to the mix. The genetic shuffling and recombination that takes place during sex affords greater variability, which allows greater flexibility in a changing world. Adaptability, pathogen resistance, and the buffering of harmful mutations are all enhanced through sexual reproduction. Instead of relying on a single genetic line, as asexual organisms do, genes provided by two parents offer a broader playing field on which evolution can run rampant.

So how does sex work? Well, the standard penis-in-the-vagina method we humans (and most mammals) employ is but one manner of achieving fertilization.

Sexual reproduction involves the union between the sex cells, or gametes, of two parent organisms. We'll start with the tried and true method of most mammals, intercourse. For intercourse to work, you need a sender and a receiver. The sender is equipped with a tool for administration (aka, the penis) and the receiver provides a welcoming receptacle (the vagina). Insertion of the tool, followed by a bit of friction, results in a magical deposit (sperm) and, if the timing is right, the sperm are greeted by an eager egg, or eggs, in many cases. From this point on, nature takes its course. The egg is fertilized and the zygote is launched on its path to personhood (or elephanthood, porcupinehood... You get the picture).

This is the typical mammalian recipe. However, nature wouldn't be nature if it didn't throw in a bit of variety. For those who lack a tool or a receptacle, or simply prefer less fuss and muss, there are other means of achieving fertilization.

For instance, take fish. Imagine a fish penis. Bet you can't. That's because most fish spawn, meaning the males and females have adopted a hands-off approach to reproduction. (I think they're on to something...) The female typically releases a cluster of eggs and the male then swims over them while discharging a milky little cloud of semen. From there, the eggs and sperm are on their own. But since there are over thirty thousand different species of fish swimming the planet, their crop-dusting method is apparently pretty effective.

Fertilization takes two main forms: internal (like us) or external (as in our fishy friends), with a few variations on these themes. And just as with most aspects of our anatomy and physiology, evolution via natural selection has tinkered, resulting in the most effective method of fertilization for each species.

Not that there aren't problems, but at least we humans are rewarded with an orgasm—or multiples, on a good day! But are we the only species that has sex for the sake of sex?

Most animals (as well as certain Republican politicians) engage in sex purely for reproductive purposes. There are just a few exceptions: humans, dolphins, and bonobos, or pigmy chimpanzees. Like us, dolphins and bonobos are known to engage in sex merely for the fun of it. Scientists believe sex among dolphins and bonobos might serve some of the same purposes it does among us humans— forging bonds which then promote group cohesion—although I'm convinced the bonobos do it for kicks and giggles. Ironically, bonobos are also one of the few animals besides us that practice face-to-face sex. Most mammals are better suited for the less personal but oh-so-effective method of mating from the rear.

You'll be happy to know that scientists are hard at work discovering all the positive effects of sex. Regular sex has been shown to improve sleep, reduce stress, increase blood flow to the brain and other organs, and even keep your ticker healthy. So, the next time you get lucky, pause for a moment and reflect on the billions of years of evolution that came before you (no pun intended).

BIRTHDAYS AND BONES
May 16

Here's a question: how many of you sneak a peek when you're in the gym locker room? Be honest. I'm one of the few who openly admits to being a gawker (of course, I do it with utmost discretion). The locker room affords the rare opportunity to see *real* bodies, not those computer enhanced, airbrushed renditions we see in the media. You get a unique double feature in the locker room: bodies of all ages and bodies that are naked (unfortunately in my locker rooms, they're also all female, but you can't win 'em all). For someone like me, who is enamored of form and function, it makes for some interesting viewing.

It's astounding how the body morphs as we age. This morning at the gym, I wrapped up my workout and was cruising to the showers when I happened by the Jacuzzi. Percolating within the bubbly froth were six elderly ladies, happily chatting away as they simmered to pruned perfection. Gazing upon that veritable senior stew got me thinking about my own aging body. Sure, the mirror is a constant reminder of the ticking gravitational clock, and our skin holds few secrets about the passing years, but how often do we contemplate the toll age takes on our skeletons?

Let's take a quick glimpse at the life of our bones. We tend to think of our skeleton as fixed and unchanging, that steely scaffold on which our flesh and muscles are suspended. But our skeleton is a highly dynamic structure, constantly changing throughout our lifetime.

Bone is made up of two primary components: the large protein, collagen, which gives the bones their elasticity; and hydroxyapatite, the dense inorganic substance that provides their strength. (Pound for pound, bone is stronger than steel!) The duality of these two

185

main ingredients is what allows bone to be strong yet somewhat flexible, and this same successful recipe is found in all mammalian skeletons.

The typical adult sports 206 bones, yet we emerge from the womb with over three hundred. Where do the others go?

They don't go anywhere. Many of the bones that make up the newborn skeleton originate in segments that fuse over time. Many baby bones also have yet to ossify, meaning they are still primarily cartilage but will eventually turn to bone (that's what makes infants so darn flexible). And some bones, such as your kneecap, won't show up until *after* you are born.

The hand is a perfect example. A newborn's hand will eventually ossify into the twenty-seven bones that form this remarkable appendage, and that, along with the twenty-six bones of each foot, make up over half of your entire skeleton!

Bones grow in two directions: they get thicker and they get longer. Length is acquired via the growth plates at the ends of bones. These plates, or epiphyses, will continue to add bone until growth ceases and the seam that separates the plate from the shaft eventually disappears. Since each bone fuses on its own schedule, they provide a handy means of determining the age of children at the time of death.

For example, say we find a skeleton whose bones are completely fused except for the collarbone. Since your collarbone is the last to fuse—usually not until you are in your early twenties—we can surmise the individual was a young adult when they died.

Just because the bones have stopped growing does not mean they remain unchanged. In fact, once your skeleton stops growing, it begins the slow march toward death. The little metabolic factories that produce bone (the osteoblasts) begin to slow down, and if they can't keep pace with the bone destroyers, or osteoclasts, bone density diminishes, and the skeleton becomes frail. That's why the elderly are prone to breaks—bone loss equals bone weakness, and fractures are the typical result ("I've fallen and I can't get up!").

There are many factors, aside from the number of candles on your cake, that age a skeleton. Our genetics can help or hinder; changes in hormones can wreak havoc on bone density; and then there's lifestyle, that most fundamental (and controllable) of causal factors. What we eat, how much we exercise, and what we smoke all affect our bones; something to think about the next time you are lying on your couch, puffing away and munching Cheetos.

So, be a good skeletal steward and take care of your bones. You only get one set per lifetime, so do everything you can to make it last. Your bones will thank you.

HOW TO CURE A CORPSE
May 23

If you've ever smelled a decaying corpse, you'll appreciate the science of embalming. Nothing smells worse than a body in the full throes of decomposition. It's a mélange of horrific odors: the gut-churning stench of putrid flesh mixed with a rank sweetness that clings to your mucous membranes. It's a fetid gift that keeps on giving.

During my tenure as a firefighter-paramedic, we were often called to "check on the well-being of a patient." In other words, "Grandpa hasn't answered his phone in over a week. Can you guys swing by and check a pulse?"

As we'd pull up to the residence, we'd scan for newspapers. An accumulation on the front lawn was always a bad sign, as was a mailbox stuffed to overflowing. We'd approach the front door and pause to take a deep whiff. We could usually diagnose death by its odoriferous calling card (or as the medical examiners refer to it, the "smell of job security").

So, if a single body is capable of producing such rank odors, imagine a battlefield strewn with dead soldiers, their bloated bodies exploding under a scorching summer sun. Those were the conditions under which embalming first took hold on a grand scale.

But first, a quick retrospective. The Egyptians were embalming over eight thousand years ago as part of their intricate mummification process. After removing the brain and internal organs, the body was immersed in natron, a salt solution that would desiccate the tissues. Once the body was dried out, which usually took about a month, the wrapping of the body would commence. Egyptians believed a person's soul would eventually return to his body, so it was only proper to preserve it for the soul's homecoming. Unfortunately, only the wealthy were afforded proper mummification; the poor, they simply dunked in salt and hoped for the best.

Fast forward thousands of years to the bloody battlefields of America's Civil War. It's the 1860s and our nation is at war with itself. Union and Confederate soldiers are busy blasting each other to smithereens, resulting in a plethora of dead bodies that have no chance of keeping fresh for the long trip home. Mass graves were the usual solution. Large pits were dug, and the bodies were laid to rest and quickly covered. But as in so many realms of science, necessity was the mother of invention.

Enter Dr. Thomas Holmes. Dr. Holmes was a captain in the Army Medical Corps, assigned to Washington, D.C. He was the first to practice embalming, which he achieved by injecting his "patients" with an arsenic solution that retarded the ravages of decomposition. His talents soon captivated President Lincoln, who was so enamored of the practice that he assigned Holmes and the Quartermaster Corps to provide embalming for Union soldiers and officers killed in battle. Holmes went on to embalm over four thousand before chucking his military career for a lucrative private practice. So, what is embalming and how is it achieved?

Today, arsenic has been replaced by a formaldehyde solution. Not only does it preserve the body, but it also disinfects. Disinfecting the corpse makes for safer handling; preserving the corpse makes it more appealing for those wishing to view the dead. It also provides a larger window in which to make funeral arrangements, freeing the undertakers from playing beat-the-clock against putrefaction.

Before the body is embalmed it is laid out, washed, and shaved, if necessary. The eyes are kept shut using eye caps, small plastic disks that are slipped under the eyelids. Perforations on the disks hold the lids in place since nothing says "creepy" like a staring corpse. The mouth is closed with tacks that are placed in the upper and lower jaws and held together with wires. A special cream is then applied to the lips to hold them in place and prevent chapping. Now the embalming can commence.

The embalming fluid enters the body via an incision into an artery, usually the carotid in the neck. A small pump holds the solution and is attached to the artery by a hose. A separate hose is inserted into a large vein, usually the femoral near the groin, and will empty into a nearby drain. The fluid (about three gallons) is pumped into the artery and circulated throughout the body, forcing out the blood and infusing the cells, before exiting through the hose in the vein. The process is complete when the entire blood volume is replaced by formaldehyde.

Idiot Alert!! There's a new trend in getting high: soaking cigarettes and joints in formaldehyde, which, when smoked, bring on hallucinations. Unfortunately, this also brings on seizures and coma, so play it safe and *don't smoke anything!*

The incisions are closed and the pump removed, but one last step remains. A hollow tube called a trocar, which is attached to a suction unit, is inserted into the abdomen. With the flip of a switch, the gasses and liquids produced during early decomposition are magically sucked away and the belly is then flushed with a preservative. Now the body is properly embalmed and the technicians, wielding their combat-ready makeup kits, can move in to restore the corpse to its former lifelike splendor.

Now you know what awaits you, should you choose the traditional route of an open casket funeral. As for me? Harvest my organs, if possible, and throw me on the barbie. I've never been one for fuss and muss.

ANCIENT AGONY
May 30

I don't know how or when she broke her leg; all I know is she miraculously survived this devastating injury. It wasn't a simple fracture of the tibia or fibula. Somehow, she sustained a midshaft femur fracture, which is not only life-threatening, but incredibly painful. It was a serious break. The broken ends of bone were knocked out of alignment, ending up side by side; the kind of fracture that can only be corrected surgically.

What force could have caused such an injury? During my years as a firefighter-paramedic, femur fractures were usually the result of an "auto vs. pedestrian" or a serious crash. It takes a lot of force to break the largest, strongest bone in the body.

But her injury was not from a car, nor from a crash, for this woman lived over seven thousand years ago along the east coast of Florida. Her skeleton was found among the ancient assemblage known as Windover, a site that has yielded a wealth of information about prehistoric Floridians.

Her bones speak of a hard life. The leg injury was but one of the maladies that plagued her and despite the fracture, she managed to survive for years afterward. The ragged bone ends eventually mended, a heavy knot of callus sealing them to each other, resulting in a shortened limb and a lifetime of limping that would have made her difficult life that much more challenging.

We can tell a lot about her through her well-preserved skeleton, buried, along with 167 others, in the base of a small pond near present-day Titusville. This mortuary pond held the bundled remains of a people who roamed the Florida peninsula thousands of years ago. Their bodies remained tucked beneath the surface until their accidental discovery in 1982.

After retiring from the Orlando Fire Department, I came to Florida State to earn a PhD, specializing in bioarchaeology and spending my years at FSU studying the skeletons from Windover. And what stories these skeletons have told.

Imagine this: It's mid-August in Florida and the temperature is hovering around ninety-five degrees. The humidity has settled like a dense fog as the sun sets on this sweltering day. The mosquitos are rising in droves from the swamp as they seek out any bit of exposed flesh, and a late thunderstorm has left behind a stifling stillness. Oh, and I almost forgot. . . you are lying on a mat, writhing in pain from a fractured femur.

Broken bones weren't the only health challenges these ancient Floridians faced. You should see their teeth.

Envision your mouth if you had never brushed or flossed (even worse, imagine your breath). But cavities and plaque were the least of their troubles. Attrition, or wearing down of the teeth, was rampant, as was tooth loss. If you were lucky enough to survive into your forties, the teeth you managed to hold onto were typically worn to the gum line from years of eating gritty, acidic foods and using your jaws as tools. Many from Windover also suffered from infections. Abscesses burrowed into the bones of their jaws, inflaming them with pus and causing full-blown sepsis if the infection entered the bloodstream.

And speaking of infection. . . their mouths weren't the only body parts affected. You may not realize it, but infection wreaks havoc on bone. It causes the outer layer of bone, the periosteum, to become inflamed and can even wheedle its way into the marrow cavity, where it can spread unchecked throughout the body.

And think how demanding their lives were. Hungry? Go hunt down your dinner or gather it from the forest. Thirsty? Trek to the nearest water source and hope it's not infected with parasites. Need shelter? Better get to work on that thatch hut, which means gathering enough palm fronds to hold off a raging thunderstorm. Need to pee or poo? Get far enough from camp to avoid contaminating the soil you'll be sleeping upon.

In other words, each of life's necessities required work. Hard work. And the result? Plenty of wear and tear on the joints. And as the arthritis advanced, each and every chore became that much harder, requiring that much more effort.

We have it made. Modern life is infused with ways to make life easier. Cars, grocery stores, washing machines, and refrigerators enable us to drive, shop, clean, and store food. Not to mention life's little luxuries, like clothes, bug spray, and toilet paper.

So, the next time you gripe about the cable being out or the Internet being too slow, take a breath and think about those in our distant past. What they would have given for one day in our shoes.

WHY WE KISS
June 6

Recently, the media was abuzz over a newly drafted NFL player. What set tongues wagging was not the four-hundred-thousand-dollar-a-year salary or the ingrained violence of pro football. Intolerants everywhere were outraged that the rookie shared a celebratory kiss with his partner, who just happened to be another man.

The squawking was immediate and profound, all of it over a simple kiss. I bet their heads exploded as they imagined the private celebration that took place later between Michael Sam and his partner, Vito Cammisano.

All the hubbub got me thinking about the significance of kissing. Why do we do it? How did it arise? What compelled our ancestors to press their lips together and go for it? Let's explore.

Kissing surely predates its appearance in the written record. I'm betting the ancients were swapping spit long before writing was invented, since Neolithic entertainment was limited to tending goats and polishing your adze. Some of the earliest references come from India. Sanskrit texts dating to over three thousand years ago describe kissing, as does the famous (and oh-so-erotic) *Kama Sutra*, although the kissing is probably thrown in to provide a break from all those sexual contortions.

The industrious Romans even devised categories to describe their kissing. A kiss on the cheek? That was called an *osculum*. A kiss on the lips? A *basium*. And deep kisses were referred to as *savolium*. Not exactly words to make you swoon but, then again, kisses were also used in their business transactions, thus the saying, "Sealed with a kiss"!

Kissing may have its origins in feeding. Many animals are known to chew food and then pass it directly to their young. Psychologists theorize that in humans, this practice may have stuck around as the child got older, morphing from a necessity to a means of bonding. Ironically, the same muscle that allows an infant to latch onto the breast also enables us to lip lock. The *orbicularis oris* runs around the outside of your mouth. Not only does it allow you to pucker up, it also allows you to contort your mouth to speak (unless you're a ventriloquist).

The *orbicularis* is only one of many muscles that engage during a kiss. The *lateral pterygoid* pulls your jaw open while the *masseter* and *temporalis* ease it shut. But some of the most important muscles for kissing reside within your tongue (why bother kissing if you don't go French?). The *styloglossus*, the *palatoglossus*, and the *hypoglossus* allow you to explore the interior regions of your partner's mouth (among other areas), but here's a word of advice to you boys: when it comes to the tongue, less is more. There's nothing worse than an overzealous "thruster," so play it cool and apply in moderation.

As we all know, the bliss of kissing is not confined to your mouth. Within seconds, your whole body jumps on board. Sensations race along the nerves of your mouth, tongue, and face, electrifying your brain, which gratefully dumps hormones and neurotransmitters into your bloodstream. Dopamine and serotonin provide that blissful rush, oxytocin intensifies your feelings of affection, and adrenaline makes your heart go pitter-patter. It's a beautiful physiological symphony. It can also be a calorie-burner. One hour of kissing burns around twenty-six calories; nothing to brag about, but it beats the treadmill.

Kissing can improve your health, along with your relationship. Regular kissing has not only been shown to increase intimacy, but it may also promote a healthy heart by reducing stress.

And we are not alone in our propensity to kiss. Apparently, our buddies, the bonobos, whose sexual exploits we explored in May's *Sex Evolves*, engage in the occasional smooch, as well.

Kissing plays a fundamental role in most cultures. According to the Bible, Judas was paid thirty pieces of silver for betraying Jesus with a kiss. The 1896 film by our own Thomas Edison, of telephone fame, was the first to premier a kiss on the wide screen in a movie blandly entitled, "The Kiss" (not very creative for a famous inventor). And a creepy side note: a print of Mick Jagger's lips sold for a whopping sixteen thousand dollars. Those smackers even sport their own Facebook page!

Your culture also dictates the context of kissing. Some countries, like China and Japan, are not big on physical contact, so public kissing is considered a no-no. In Western Europe, kissing has become the normal part of a meet-and-greet. And in Muslim countries, kissing is reserved for those of the same gender, which is practically unheard of in the good ole U.S. of A. Men kissing men?? You might as well ask them to put on a pair of pantyhose... which is really a shame. One of the things I loved about traveling through Italy was seeing men greet each other with a kiss to the cheek. So refreshing.

And apparently there are many different ways to kiss. The most expansive website I found boasted fifty different techniques, from the "freeze kiss," where you add a piece of ice to the mix, to the "buzzing kiss," which entails vibrating your lips and cheeks while humming next to your partner's face. The ice, I can handle, but any "buzzing" partner of mine is going to get the fly-swatter.

A kiss is indelibly intimate. Sex itself can be an impersonal affair—ask any prostitute—but there's something about a kiss that transcends the physical boundaries that separate us. So, pucker up and enjoy the rush!

UP FROM THE BOGS
June 13

My friends know me well. They are aware of my fascination with all things dead, thus my office is littered with their ghoulish gifts. There are the numerous objects sporting skeletons: from coffee mugs, to post cards, to bone-shaped candles. And then there are the actual dead critters that line my shelves: the beautiful skulls of alligator, raccoon, and turtle, along with my latest acquisition from the leftovers of a physician's anatomical collection—a bowl made from a human skull set atop a pelvic bone. It's a beauty.

But some of my favorite oddities are found on my windowsill. There reside the corpses of a giant spider, a hairy moth, and the crème de la crème, my mummified lizard collection. The lizards had the misfortune of becoming trapped in our office basement. They were fairly fresh when found and I had the joy of watching them mummify over the accompanying weeks. Now their desiccated little bodies adorn my windowsill and keep me company throughout my busy day.

Mummies are cool. We're all familiar with the classic Egyptian mummies: those meticulously tended bodies surrounded by a litter of elaborate grave goods. But what I find even more fascinating are the mummies produced by nature. I'm talkin' bog bodies.

As a bioarchaeologist, I've spent years studying the skeletons of ancient Floridians. Most of my research has focused on the remains from Windover, a seven-thousand-year-old pond cemetery from which 168 well-preserved skeletons were excavated. (See April's *Ancient Agony*). The pond provided the perfect environment for the preservation of bones: a neutral pH and a thick, anaerobic layer of peat in which the bodies were carefully placed.

Bog bodies are from similar settings. They too come from "wet sites," yet the peat bogs that have produced these bodies, namely sites throughout northern Europe, tend to be acidic due to the presence of Sphagnum moss. The moss prevents bacteria from decomposing soft tissues, leaving behind skin, internal organs, and clothing, and providing virtual windows into the past.

The bogs are not formal cemeteries. Most appear to be the dumping grounds for those who met violent ends. Because of the remarkable preservation, many still exhibit the tools of their demise. Leather thongs used for asphyxiation, garrotes for strangulation, and ropes for hanging are recovered, still encircling the necks of their ancient victims. And when I say ancient, I mean ancient—some even date back ten thousand years. Let's take a closer look at a few of these amazing bodies.

Way back in 1879, the remains of a woman were discovered in a bog in Denmark. She apparently died over two thousand years ago, her body marred by repeated hacking wounds, her right arm torn from her body. A similar fate awaited a sixteen-year-old girl who was strangled and dumped in a bog in Holland. She was recovered in 1897, still wrapped in her worn, woolen cloak, the cloth used to strangle her still bunched around her throat.

Straddling the borders of western Germany and the Netherlands is Bourtanger Moor. Today, the moor is a vast nature preserve, but back in 1904, the bog produced the shriveled remains of two men, found side by side. We don't know how one of the men died, but the other suffered a stab wound large enough to cause his intestines to escape up through his chest. Not a happy ending for this duo.

But one of the most famous (and well-preserved) of all bog bodies was discovered in 1950 by a peat-cutting crew near the village of Tollund in Denmark. Tollund Man, as he came to be known, was a middle-aged man, just over five feet tall, who died around 400 B.C., apparently from strangulation. The four-foot-long braided leather strap used to choke him was left in place; faint indentions are still visible in the skin around his neck. He sports a small cap made from sheepskin, worn fur-side-in, probably to keep his head warm through the frigid Denmark winters. Two delicate leather straps hang from each side of the cap and you can just picture him securing them beneath his chin on cold, windy days. A leather belt still adorns his waist; the rest of his clothes were dissolved within the bog.

His face is extremely lifelike, as if he had simply stopped for a quick nap. His eyes are closed, his face relaxed. Even the stubble on his chin is intact and if you stare at him long enough, you can almost imagine his eyes fluttering open as he awakens from his two-thousand-year siesta. The excellent preservation extends beyond the superficial. His stomach and intestines provided clues to his last meal, which took place about twelve hours prior to his death. Thirty different types of plants were identified, including oats, barley, and flax. Two adventuresome archaeologists in 1954 took it upon themselves to recreate Tollund Man's last meal, but their efforts were rewarded with a dish described simply as "dreadful." Gotta give them credit for trying. . .

And Tollund's face isn't the only body part that is beautifully preserved. His feet appear to be perfect wax-like renditions, his toenails still visible. He spent much of his life barefoot. The pads of his feet are well worn and show scars from previous injuries. His body is tucked on its left side, his legs curled, arms bent. Parts of his skeleton remain, the bones protruding along his extremities, his ribs visible beneath his glossy flesh.

Archaeology, and more specifically, bioarchaeology, can reveal much about the past through the skeletons left behind. And when those skeletons happen to remain fleshed, they provide a more intimate glimpse of ancient life. To look into the face of an individual who lived thousands of years ago, before cars and planes and computers, is a truly wondrous experience. So, the next time you are traipsing through a bog, be on the lookout. You just might stumble upon the past.

EVOLUTION OF THE BRA
June 20

This morning while I was getting dressed, I realized that throughout my entire adult life, I've worn the same type of bra. It's really a grownup version of a training bra: stretchy material that comfortably conforms to that most difficult of female physiques—the broad-shouldered, small-breasted tomboy variety.

For years, I've pumped and swum my way to a sturdy, muscular frame. Of course, much of that was dictated by the thirteen years I spent as a firefighter-paramedic, where I hoisted myriad tools, equipment, and patients on a regular basis. And even having left the field of emergency services, I still adhere to a rigorous workout regime.

But all those years of working out did little for the development of my female upper regions. Alas, I still sport the chest of a prepubescent teen and shall forever be relegated to the training bra.

Bras have truly evolved over the centuries. They reflect fashion, function, and sexuality and are designed to lift, support, and separate those most essential female accoutrements, the boobies. So, I thought it would be fun to travel back in time and witness the evolution of this imperative undergarment.

Bras were believed to have evolved over the last hundred years: a result of women kicking their corsets to the curb after centuries of miserable cinching. But in 2008, a couple of archaeologists stumbled upon four linen brassieres tucked within an Austrian castle dating back to the 1400s, pushing the invention of the bra back another five hundred years. These ancient undergarments resemble their modern counterparts, some even sporting intricate decorations of lace. They are tattered and torn, calling to mind men's jockey shorts, which are traditionally worn until they disintegrate.

In 1913, a corseted Mary Jacob was wrestling her way into a frock when she decided there had to be a better way. Using a ribbon and two handkerchiefs, she constructed a rudimentary bra (use your imagination), squishing and flattening her breasts into what is described in the literature as a "monobosom" (again, use your imagination). Word of her handy-dandy invention quickly spread, so a year later she patented her creation and capitalized on her busty achievement.

Around this time, World War I was winding down and the corset was on its way out. The metal that lined those miserable contraptions was needed for ammunition and tanks, so women everywhere gratefully complied. Styles were changing, too. The curvaceous figures of the corseted era were being replaced by the boyish, flat-chested flappers of the Jazz Age (damn, I would have been popular). Women were chopping their hair and shortening their skirts. They didn't have time to mess with corsets. Besides, when you are up all night drinking bootleg whisky and bebopping to Louis Armstrong, breasts just get in the way.

In 1922, in a small dress shop in New York, Enid Bissett, along with Ida and William Rosenthal, were hard at work perfecting what would become the modern bra. They devised a two-cup device that was held together with elastic and sewn into the dresses they designed. They soon realized they were sitting on a gold mine, so they stripped the bras from their dresses, sold them outright, and the Maidenform Bra was born! Since women come in all shapes and sizes, they devised the alphabetical designation for cups. The letter is based on how many ounces of booby the cup will hold: "A," about eight ounces; "B," thirteen; "C" can contain twenty-one bodacious ounces; and the daddy of all cups, "D," can hold a whopping twenty-seven. That's a lot of breast! I wonder where my bras rank. I'm thinking somewhere in the tablespoon realm.

An interesting side note: It was the American author, Mark Twain, of *Huckleberry Finn* fame, who devised the oh-so-clever elastic strap and metal clasps. He envisioned applying his elastic-clasp combo to an array of undergarments, but the pantaloons industry was in its death throes, so he stuck to bras. But I digress.

About forty years later, human ingenuity took another giant leap forward. No, I'm not talking about the moon landing, I'm talking about the invention of the Wonderbra. With streamlined cups and a plethora of padding, even the "A"-leaguers could sport a bit of cleavage. But the real breakthrough came in 1977 when Roy Raymond founded Victoria's Secret. Although Frederick's of Hollywood had been around since the 1940s, many were too embarrassed to venture inside. Victoria's Secret provided a safe haven for the bashful, and their catalogs, masturbation fodder for teenage boys everywhere.

But have you every stopped to ask yourself why?? Why do women wear bras? And more importantly, why do we consider "the natural look" obscene? Yes, bras hold everything in place, which becomes more important as the years tick by, but it's disturbing that something as natural as unfettered breasts is deemed lewd and indecent.

I'm sure the menfolk would be in favor of a braless revolution. As for me, I'd participate, but I doubt anyone would notice. . .

WHY WE STINK!
June 27

Humans can be a stinky species. Let's face it, our bodies are quite adept at producing odors, many of them fairly rank. From orifices to crooks and crevices, our bodies give off a range of scents that can titillate or torment, depending upon conditions.

One of the most odoriferous places I've ever encountered is the London Underground. During the summer of 2002, I was completing my museum internship at the Natural History Museum and was a frequent traveler aboard the Tube. And because London is such a multicultural city, its public transportation provides a vast range of bodily aromas. B.O. abounded, as did the smell of unwashed hair and feet. Cooking spices lingered on breath and skin, intensified by the stifling heat of the subway, making it feel as if we were trapped in a steaming pot of tandoori. When the train finally screeched to a stop and the doors flew open, fresh air would rush in as the aromatic mob rushed out.

We've already covered the intricate dance between nose and brain in October's *Something Smells*, so I thought we would explore our body's amazing ability to produce odor. Let's start at the head and work our way down.

We've all experienced bad breath (either personally or in our partner), especially in the morning, when the bacteria that infiltrate our mouths have had hours to multiply unchecked. Those bacteria, if left to their own devices, can cause gum disease, which in turn can cause chronic bad breath, technically known as halitosis. But gum disease is only one cause of a stinky mouth. What you eat can also affect your breath, especially foods that stink on their own. Onion and garlic not only pass through your mouth and down your gullet, but they are also absorbed by your bloodstream where they travel to your lungs and are eventually exhaled. Not a good scenario if intimacy is on the agenda.

Other harbingers of halitosis include conditions such as diabetes and acid reflux, poor fitting dentures that trap food, and the worst offense of all—*smoking!!* Not only are you destroying your lungs, but you are also destroying the nasal passages of those you kiss. There's nothing worse than French kissing an ashtray.

Moving down the body, we arrive at those classic stinkers, the armpits. Your body has two types of sweat glands which serve two different functions: the eccrine glands, which cover most of your body, and the apocrine glands that reside beneath the hairier regions, such as the armpits and groin. As your body heats up, the eccrines cool you down by secreting fluid onto the surface of your skin. The apocrines, however, are triggered by stress and respond by producing a milky fluid into your hair follicles. Although the fluid is odorless, it blends with bacteria on your skin to produce that common and oh-so-pungent aroma, body odor.

Moving south, we arrive at the genitals. Here, our paths diverge, for odor in the nether regions depends on whether you sport a penis or a vagina. They each have their own aromatic obstacles.

Women are all too familiar with issues of odor, for whenever you possess a warm, moist, cavity, sooner or later, it's going to stink. Normal sweating and fluctuations in your menstrual cycle can cause odor, but if it persists, bacteria are probably to blame. The most common cause of vaginal odor is a lovely condition known as bacterial vaginosis—an uprising of the normally occurring bacteria in which they overrun the V, leading to a smelly situation. Poor hygiene can also cause a stench, and many women respond by douching 'til the cows come home. Big mistake. Douching can lead to other problems by further upsetting the bacterial tug-of-war. A bit of advice: if you smell like the docks, it's time to call your doc.

And although you guys are vagina-free, it doesn't mean you can pass the smell test. Remember our buddies, the apocrines? They kick into overdrive when you are emotionally stressed (or excited!) and the furrier you are down there, the more apt you are to smell. What do the experts recommend? You can bump up your personal hygiene routine, wear loose clothing (preferably cotton), and whip out the hedge clippers to keep things tidy.

We finally arrive at the feet. Yes, foot odor is a common malady and even has a super-scientific name: *bromodosis*. Feet stink because feet sweat. In fact, there are more sweat glands on our feet than anywhere else on our bodies! And the more we sweat, the more we stink. To make matters worse, some folks suffer from hyperhidrosis, aka, excessive sweating.

But just like our smelly mouths, pits, and genitalia, there are remedies for smelly feet. Changing your shoes, wearing cotton socks, and using powder to absorb moisture can thwart foot odor. In fact, all of these stinky situations can be remedied with proper hygiene.

Your body is a complex organism, fueled by millions of chemical reactions taking place as you eat, sleep, and go about your daily routine, so it's no surprise that some of those reactions produce funky smells.

So, brush your teeth, take a shower, trim your hairy regions, and change your shoes, and you too can be odor free. Your body—and those around you—will thank you.

TALES FROM THE POO
July 4

They say big things come in small packages. This is certainly the case with archaeology. We can take the smallest bit of evidence—a fragment of bone, a shard of pottery—and extract a wealth of information from the pieces of human history buried within the archaeological record. And one of these small parcels of the past offers an array of cultural, biological, and environmental information. I'm referring to ancient poo.

Since we've already explored the myriad ways humans emit odor (last week's *Why We Stink!*), I felt it only natural to highlight one of our smellier byproducts, feces. And from the standpoint of a bioarchaeologist, who relies on ancient skeletons to piece together history's complex puzzle, I've developed a respect and admiration for the excrement of yore. So, let's take a look at these piles from the past, otherwise known as "paleofeces."

Everything we consume traverses the long and winding road of our GI tract, providing nutrients and energy along the way before being deposited as waste. Just as people poo today, they were pooing in the past, and, occasionally, a lucky archaeologist will stumble upon a pile. Lo and behold, a window to the past is thrown open.

Paleofeces are commonly called "coprolites," a term coined by geologist William Buckland back in the 19th century as he led the charge in dinosaur discoveries. He not only discovered dinosaur bones, but he also identified their monstrous mounds of manure and today, paleontologists around the world devote whole careers to the study of dino dung. But we're going to home in on the human version of these exotic treasure troves and explore the variety of information gleaned from ancient turds.

I bet you're wondering how a pile of poo can be preserved for thousands, or even millions, of years. Well, like any bit of archaeological data, the conditions must be ripe for preservation. In the case of paleofeces, they manage to survive when they become desiccated (dried out) or fossilized (turned to stone). As the poo dries and hardens, its contents become fixed. The poo must then be reconstituted, or restored to its original consistency, which is done using special solutions that help preserve what's tucked inside. Interestingly, as the poo is reconstituted, not only does it offer up its contents, but it also emits its original odor! Once the poo is soft and pliable, the investigation can begin.

For anyone who has ever eaten whole-kernel corn, you know that the kernels magically reappear in your excrement (my fellow firefighters referred to them as "tracers"). The study of paleofeces works on this same principle, for ancient poo reveals a lot about the foods that were consumed, especially plants. Pollen helps us identify the plant species, as do the phytoliths, the portion of the plant composed of silica, which resists decay (and digestion). And why are plants important? Plant consumption not only tells us about diet (what people were eating), and environment (what conditions they experienced), it also helps us reconstruct when domestication took place in different parts of the world. By identifying the plants and then dating the poo, we can obtain the "when" and "where" of agriculture.

But plants aren't the only things recovered from paleofeces. In many cases, critters are lurking within, mainly in the form of parasites. Parasites are organisms that take up residence on or within a host, drawing nutrients (at the host's expense) to stay alive and reproduce. Since many parasites occupy the gastrointestinal tract, they are passed out of the body during defecation where they can then be spread to new hosts, which is why hand washing is so vital. When parasites are discovered in paleofeces, they provide information about health and nutrition among ancient populations (it's hard to be healthy with a belly full of worms). The study of ancient parasites has become so popular it has now evolved into its own specialty: paleoparasitology.

Along with plants and parasites, we also find the bones of animals, and even DNA, when we are lucky. So where do we find these valuable little nuggets? They show up in the ground, in ancient latrines, in mummified corpses, and even in the bog bodies of northern Europe. And they are providing a vast array of information about our past.

Evidence for people in North America was pushed back a few thousand years when a fourteen-thousand-year-old pile of poo was discovered in Oregon. And the earliest evidence for the domestication of dogs came from human coprolites in Texas, dating to over nine thousand years ago. Unfortunately, the dog ended up on the menu, but it's still a cool discovery. And in a bit of poetic justice for critters, the oldest known human hair was recently discovered in, of all things, a crusty little hyena turd recovered from a cave in South Africa. All that from simple dung.

But you don't have to be an archaeologist to appreciate the value and beauty of ancient poo. Thanks to some inventive jewelry designers, you can now adorn yourself or your loved one with real live dino droppings. So, the next time that birthday or anniversary rolls around, say "I love you" with a bit of biological bling. Who says you can't polish a turd?

THE BODY AS EVIDENCE
July 11

Although all humans are of the same species, with the same overall body parts, there are a number of attributes that are specific to each individual, even in the case of identical twins. These specifics come in handy when it comes to solving crimes.

During grad school, I had the good fortune of completing a Directed Independent Study at the C. A. Pound Human Identification Lab, located on the outskirts of the University of Florida campus. I trained under the assistant director, Dr. Michael Warren, a gifted forensics expert who would go on to play a leading role in the infamous Casey Anthony murder trial. The training was meant to supplement my studies in bioarchaeology, for both forensic anthropology and bioarchaeology utilize many of the same techniques. Both specialties rely on the examination of the skeleton; however, bioarchaeology deals with ancient remains while forensics focuses on contemporary deaths, typically those occurring within the past fifty years.

Working at the Pound Lab was a great experience. Each day, Michael would pull a box from the back shelves where the old cases reside and I would set to work on the remains, placing the bones in anatomical order, determining the age, sex, and height of the person, and noting any pathologies and unique identifiers, such as dental work or skeletal anomalies. From X-raying decapitated heads to defleshing a young girl, my work at the lab provided an up-close-and-personal glimpse of forensic science (along with a few cases of the dry heaves).

So, I thought we would explore how our bodies serve as evidence—not as victims, but as perpetrators. Let's examine the clues our bodies leave behind.

Say you commit a heinous crime, one that includes the deadly duo of rape and murder. There are myriad ways you can be linked to that crime. Let's start with the tried-and-true method, fingerprints.

No two people share the same fingerprint. Even identical twins, who have the same DNA, will have differences in their fingerprints, since the "friction ridge skin," as it is known in professional circles, is a result of many environmental factors, such as bone growth and conditions within the womb. And although Sir William James Herschel is credited with being the first to use fingerprints as a means of identification back in 1858, it seems the Chinese were about two thousand years ahead of the curve.

In "The Volume of Crime Scene Investigation-Burglary," dated to around 200 B.C., the Chinese mention the use of handprints as evidence. They were also using fingerprints on clay seals to secure documents, which is pretty clever, considering they wouldn't invent paper for another three hundred years.

It wasn't until 1892 that fingerprints were first used to solve a homicide. That accomplishment goes to Argentina, whose Buenos Aires sleuths relied on prints to solve the Rojas murder case in which the mother of two brutally slain boys eventually confessed after her bloody fingerprint was identified at the crime scene. But fingerprints are just the start of the biological trail. Let's talk semen.

There are usually about two hundred million sperm swimming within a single ejaculate, making it the perfect fluid for DNA analysis. Even individuals who can't produce sperm (aspermiacs) and those who have been clipped (vasectomized) can still be identified based on their seminal fluid. The test for semen at a crime scene is cheerfully known as "The Christmas Tree Stain," since the reagents stain the sperm in blues, greens, and reds. Once the sperm are recovered, the DNA packed within each little swimmer can be matched to a culprit.

But semen is just one of several bodily fluids used in identification. Blood is another popular means of narrowing down, or eliminating, suspects. By identifying the grouping (A, B, O, or AB), the presence of Rhesus antigens (Rh factor), and certain genes, investigators can match or disqualify an individual based on their results.

Saliva is another common fluid recovered at crime scenes. It can be found on the victim, as can bite marks, or left on objects such as cigarette butts, glasses, or cans of soda, and is another valuable source of DNA. But bodily fluids, which fall under the specialty of "forensic serology," also include anything excreted or secreted from the body, including urine, vomit, oils from the skin, and feces.

And let's not forget about hair. The human body sports around five million hair follicles. On average, we lose around one hundred hairs per day. The length and quality of the hair is based on the region of the body from which it sprouts, and ancestry dictates whether it is light, dark, straight, or kinky. Microscopic characteristics of hair enable investigators to differentiate between individuals and to determine whether the hair has been chemically altered, thus providing another link between suspect and crime scene.

Whether it's semen, blood, spit, or hair, our bodies, as living organisms, leave behind a biological trail. And as the science of forensics becomes more exact, that trail becomes easier to follow. Something to think about the next time murder crosses your mind.

BODIES IN MOTION
July 18

This month I resigned from my job. Not because it's a bad job; I resigned based on a visceral feeling that it's time to move on. It's not the first time I've walked away from my life. After thirteen years as a firefighter-paramedic, I experienced the same sense of restlessness; that consuming need for a new direction. So, I've cleaned out my office, sold my house, and will now venture forth into the unknown.

Fortunately, I'm not without means. I have my fire department pension and a nice little nest egg put aside. A life-threatening allergy to small children has enabled me to skirt parenthood, so I'm in pretty good shape financially. Therefore, I'm chucking this life to do a bit of wandering and see what pops up on the horizon.

Moving is always a pain. I'm determined to scale back my life and rid myself of unnecessary baggage: stuff I've accumulated over the years that has lain dormant and untouched. I've even started parceling off my library, the ultimate sacrifice for any self-respecting nerd.

My impending move got me thinking about movement in general. So, to divert my thoughts from packing (aka, procrastinate), I want to discuss the amazing ways our bodies move us through life.

Imagine reaching for a cup of coffee (or in my case, a glass of gin), taking a bite of a sandwich, or brushing your teeth. These actions—reaching, lifting, connecting, and returning—are accomplished with barely a thought. We take for granted that our bones and muscles will work in concert to achieve the desired action, which is all the more apparent when we suffer an injury that limits movement.

The 206 bones in your body and the six-hundred-plus muscles on your frame (known collectively as the musculoskeletal system) cooperate to move your body through space. Since contracting muscles can only pull on the skeleton, there must be opposing muscles that reverse the action. For example, when you bend your arm to take a bite of that sandwich, your biceps flexes while your triceps relaxes. When you straighten your arm to put the sandwich down, they reverse roles. This "working together" is referred to as antagonistic action and it occurs throughout the moveable parts of your body.

There are various types of body movement. Flexion is just as it sounds: bending at a joint, as when you bring your foot up to the back of your thigh. Extension is the opposite: a straightening of the joint as you return your foot to the ground. The movement of your limbs away from your body is known as abduction (as when you reach to hang something up); adduction involves moving them towards your midline. And finally, circumduction is the movement of a limb around a joint, as when you swing your arm in a wide circle.

But these intentional movements don't just happen, even though many of them seem automatic. Movement originates in the central nervous system, comprised of the brain and spinal cord, where lightning-fast impulses race along motor neurons. These impulses cause your body to react via the peripheral nervous system, which includes your arms, hands, legs, and feet. Sensory nerves deliver information back to your brain and it's this beautiful synchronization between motor and sensation that enables movement. Think about it. . . every single movement you make, from the smallest (batting your eyes) to the grandest (jumping a puddle) is carried out through the communication between brain, nerves, bones, and muscles. It's truly mind-boggling.

But what about unconscious movement? Even if we sit perfectly still, parts of our bodies are still moving, acting on directives from the brain that never register consciously. And thank goodness. Imagine if you had to direct your heart to beat, your lungs to expand, your bowels to digest, eyes to blink, or your throat to swallow. You probably wouldn't get a whole lot done. That's the beauty of the autonomic nervous system, that complex arrangement of nerves that connects the central nervous system to the heart, lungs, and many of your internal organs. It works reflexively and involuntarily, maintaining a steady state (homeostasis) within the body, controlling heart rate, breathing, and circulation, along with all the other bodily functions that keep us alive.

The body's movements are wondrous to behold. They keep us alive, physiologically and emotionally. As we run or dance or embrace, our hearts are beating, our blood is coursing, and our breath is rushing in and out. Imagine your life without movement. Think about those who lose their ability through accident or disease; those stoic individuals who live on in the face of paralysis. They are truly the face of courage.

So, as you move through your day, think about this wonderful gift, and take none of it for granted. Movement propels us forward and is graciously complemented by the leaps and bounds of our minds.

RESUSCITATING THE DEAD
August 1

How many of you have ever seen CPR performed on a victim? Better yet, how many of you have performed it yourself? If you've never seen or done it in person, I'm sure you've witnessed the hokey representations on TV, where the medical crews pump on the victim's chest for a few minutes and are rewarded when the patient sits up and says "howdy." Unfortunately, it doesn't really work that way.

This week, as part of my paramedic renewal, I completed my recertification in Advanced Cardiac Life Support (ACLS). ACLS provides guidelines for cardiac emergencies, including CPR, drugs, and advanced techniques such as airway procedures and defibrillation. The certification must be renewed every two years, so that providers remain proficient in their skills and informed of guideline changes.

I've been cramming for the past week, studying EKG strips, calculating drug dosages, and memorizing the latest protocols in an effort to avoid embarrassing myself during my exams. All that studying got me thinking about the history of cardiopulmonary resuscitation (CPR) and how far we've come in our attempts to revive the dead. So, let's take a peek at this convoluted history.

First, for those unfamiliar with the basics of CPR, let me give you a brief overview. If someone goes into cardiac arrest, CPR, when performed properly, provides breathing and circulation for the downed victim. Compressions to the chest circulate the blood, thereby oxygenating the tissues, and ventilations done mouth-to-mouth (or better yet, mouth-to-mask) provide the necessary oxygen. These measures, performed as soon as the victim collapses, greatly improve his chance of being revived, since the longer one goes without circulation, the less likely one is to respond to advanced treatments.

Early medical practitioners understood the basics, even though it took them a while to put the pieces together in a coherent strategy. As far back as the 1500s, the physician and anatomist, Andreas Vesalius, realized that animals could be kept alive by forcing air into their lungs. The alchemist, Paracelsus, tried resuscitating a corpse using bellows, a technique he lifted from ancient Arabic medical scripts. And there are even references to mouth-to-mouth in the Bible. Elisha is reported to have breathed into a child to revive him, so apparently people have been dabbling in artificial respirations for quite some time.

It wasn't until the 1740s that the Paris Academy of Sciences publicly recommended mouth-to-mouth for drowning victims. But following the discovery of oxygen in the 1770s, practitioners decided the exhaled air delivered via mouth-to-mouth was too deoxygenated to be of much use. Mechanical methods had also fallen out of favor after a few mishaps with the bellows (apparently, it's pretty easy to blow out someone's lungs if you're not sufficiently trained), so the idea of artificial respirations fell by the wayside and the focus shifted to chest compressions.

It was well understood by this time that blood circulated throughout the body and that circulation was fundamental to life (battlefield injuries probably provided the obvious link between bleeding and rapid death). A Mr. Moritz Schiff, in 1874, noted that pumping the chest of a dead dog produced a carotid pulse, and word of his canine cardiology quickly spread. It was soon tried out on felines when Rudolph Boehm and Louis Mickwitz successfully resuscitated a kitty by squeezing its chest and ribs. But it wasn't until the 1890s that Friedrich Maass actually applied it to a human.

Unfortunately, the concept of cardiac massage would also fall by the wayside for the next seventy years, until 1958, when William

Kouwenhoven reintroduced chest compressions in treating the pulseless.

Despite the evolving science, misconceptions about reviving the dead persisted. Folks believed the stricken merely required the proper stimulation. Bodies were placed in rolling barrels or strapped to horses in hopes of jostling them back to life. A naïve but industrious sector even devised crazy contraptions on which the dead could be resurrected, and these methods hung on well into modern times.

Through trial and error, it was finally established that a combination of ventilations and chest compressions could provide adequate respirations and circulation for victims of cardiac arrest. Lifesaving took another giant leap forward with the discovery of electricity and its effect on the heart. Since the heart relies on electrical impulses to generate a pulse, and certain types of cardiac arrest are due to erratic electrical impulses (known as fibrillation), researchers devised a method of delivering a shock to the heart that could correct the dysrhythmia. Thus, the tried-and-true technique of cardiac defibrillation was born.

Today, millions are trained in CPR, professionals, and laypeople alike. And it's a good thing, for immediate CPR is an essential link in the "Chain of Survival." By recognizing an unresponsive person, activating emergency medical services, and quickly initiating CPR, you too might help save a life. So, get trained and stay current. The next victim could be someone you love.

UNINTELLIGENT DESIGN
August 8

Last month, doctors removed an astounding 232 teeth from the jaw of a young boy in India. The teeth, which ranged in size from small flecks to marbles, were the result of a rare disorder called "complex composite odontoma" and were removed using a hammer and chisel, since the size of the growth precluded simple extraction. When all the fragments were removed (along with a normal molar affected by the tumor), the boy's jaw was sewn up and he went happily on his way.

By normal standards, this is a strange case, but these peculiar tales crop up on a fairly regular basis. It seems the human body is always devising bizarre ways of expressing itself, like the eighty-seven-year-old woman from Switzerland whose esophagus would twist into a corkscrew every time she ate, or the nineteen-year-old Iranian boy who had a tumor removed from his eye when it started sprouting hair—an apparently normal response for limbal dermoid tumors, which have also been known to grow cartilage and sweat glands.

These stories represent odd conditions in which the normal structures of the body have gone awry. But you don't have to search far and wide for biological oddities. Our bodies contain a number of strange and nonsensical structures that no intelligent designer would include when drafting the model *Homo sapiens*, and I have spent considerable time imagining alternative arrangements, some of which I'll share with you today. So, let's take a look at a few of these anatomical anomalies.

Let's start with our eyes. There are numerous problems with the design of the eye, which is ironic since many creationists use the intricate construction of the eye as proof of a creator. If that were the case, why would our retinas be inside out? Why would our photoreceptor cells (the rods and cones) be displaced by the optic nerve and vessels, resulting in a "blind spot"? And why the hell do so many of us have to wear glasses?? If only we had the eyes of a gecko. Nocturnal geckos sport zig-zagged pupils that make for excellent night vision, and their little wandering orbs are about 350 times stronger than ours.

And what about our throats? It's a sad state of affairs that our esophagus and trachea originate in our mouths. That small flap of cartilage known as the epiglottis, whose sole purpose is to slam shut to prevent food and drink from "going down the wrong pipe," fails on a regular basis, resulting in many a fatal choking. A better design would be two separate openings: one for air and one for food. If only we had a blowhole, we could breathe and swallow in unison. No wonder dolphins are always smiling.

Since we started our story with teeth, let's address the problem of dental crowding. Most humans can no longer accommodate the full set of thirty-two, forcing many of us to have our wisdom teeth yanked. We'd be better off if we simply did away with the third molars. We can obviously survive just fine without them. Despite having mine removed, I've gone on to lead a full and healthy life yet was relegated to two torturous years of braces to correct the crowding they left behind.

And how about our nipples? How did they end up so high on the chest? They would be far more practical at waist level, allowing mothers to simply nurse their infants from their laps. And why do men even have them? Aside from a tickle during foreplay, they really serve no purpose on you boys.

And speaking of sex. . . vaginas are another issue, especially when it comes to childbirth. Because of the route the vagina takes through the base of the pelvis, our bigheaded infants must pass through the tight confines of the pelvic outlet. This arrangement can make childbirth a dangerous venture. We'd all be much safer if the newborn was simply expelled via the umbilicus. Or better yet, if we could transform ourselves into marsupials and carry the little bugger around in a pouch. When it was ready to emerge, it could simply scuttle up to the nipple, and if they were waist high, our little joey wouldn't have to scuttle nearly as far!

The human body is full of idiosyncrasies, proof that we did not arrive fully formed. Nor are we models of perfection constructed by an omniscient designer. Our bodies are the result of millions of years' worth of evolutionary tinkering. Natural selection has tweaked our structures to provide greater adaptability and an evolutionary edge, yet our flaws are numerous, our quirks abundant.

So, embrace your oddities and keep in mind they are part of our complex evolutionary legacy. Besides, perfection is overrated.

UP IN SMOKE: A BRIEF HISTORY OF CREMATION
August 15

Around forty thousand years ago, along the shores of Lake Mungo in southeastern Australia, a small group of aborigines took time out from wallaby hunting and emu egg gathering to cremate the remains of a woman. Mungo Woman, as she is affectionately known today, was cremated and her bones crushed, then cremated again, before being buried along the shores of this ancient lake. Discovered in 1968 by a young geologist scouting for fossils, Mungo Woman would achieve fame not only as one of the most ancient skeletons from the land down under, but also the oldest human cremation yet discovered.

It took a while for cremation to catch on. The tried-and-true method of earthly interment began around one hundred thousand years ago and, although our friends in the Outback may have gotten a jump start on cremating, it wasn't until around five thousand years ago that it took hold as a custom. So, let's trace its origins and see how it's become one of the most common methods of disposing of the dead.

Cremation on a regular basis first crops up in northern Europe and the Near East. Stone Age Slavs were producing decorative urns in which to house their loved ones' remains and the practice soon spread into the British Isles and southern Europe. By three thousand years ago, the Greeks had jumped aboard the cremation train and it quickly became their primary means of disposal. Cremation was quick and clean, and proved a handy means of disposing of the dead, especially during times of war (aka, most of Greek history).

By the rise of the Roman Empire, around two thousand years ago, cremation had taken hold in Italy (an excellent means of dispensing with dead gladiators). Urns had become so elaborate that they

222

required special repositories; they were simply too pretty to stick in the ground.

But with the rise of Christianity, cremation found itself in a death spiral. Christians shunned the practice since it smacked of paganism, and it soon fell by the wayside, except during the rare instances of plague and war. From this point on, the dead were placed in the ground and it would take another fifteen hundred years for cremation to emerge from the ashes (couldn't help myself).

And emerge it certainly did. But it didn't just slink back onto the stage. Modern cremation sprang to life at the opening of the Vienna Exposition in 1873, when an Italian by the name of Brunetti displayed his cutting-edge oven, guaranteed to reduce your loved one to a fine, ashy powder! A year later, in England, Sir Henry Thompson, surgeon to the Queen and enamored of the thrift of cremation, founded the Cremation Society of England, and crematoriums began cropping up in England and Germany.

Never to be outdone by our friends across the pond, Americans began experimenting with cremation—which is ironic, since Native Americans had been cremating their dead for thousands of years before the palefaces came ashore. The first American crematorium was built by Dr. Francis LeMoyne in 1876 on a small parcel of his land, aptly named Gallows Hill, within the quaint community of Washington, Pennsylvania. Fearing the town's crowded cemetery was leeching into local water sources, Dr. LeMoyne deemed cremation a more sanitary alternative, so he spent fifteen hundred dollars of his own money to have the small brick building erected.

He designed the oven to prevent direct flame contact with the bodies, and the first cremation was lit off on December 6th. Sadly, of the forty-two cremations that took place at Dr. LeMoyne's crematorium, his would be number three. He was slid into the oven in 1879 and his ashes were placed in a small urn in front of the building, which still stands today and is run by the Washington County Historical Society (call ahead for private tours).

Today, cremation is all the rage. It is a world-wide custom practiced in most societies and integral to many religions. From Buddhists to Christians, from Hare Krishna to Quakers, burying ashes has become an accepted method of interment, although many have specific guidelines. Yes, there are religions that shun the practice— Islam, Orthodox Judaism, and even the Presbyterians, who don't forbid cremation, but simply prefer the ground—but cremation provides an economical alternative to the rising (and ridiculous) cost of traditional burial.

In fact, a quick Google search in my area not only turned up a plethora of crematoriums, but they also sport catchy names. According to their website, Island Cremations, located on Merritt Island, provides an all-inclusive deal for the bargain price of $695. (*"Don't overpay on your most difficult day!"*) There are even crematoriums for your pets! Pet Passages, which claims to be "The Leading Authority in Pet Loss," has a website where you can choose the type of service, pick from a beautiful collection of stylistic urns which start at seventy dollars, and peruse a list of books to help get you through your loss (such as *Cold Noses at the Pearly Gates*, by Gary Kurz).

So, if you haven't done so already, give cremation some consideration. According to the fun folks at the National Funeral Directors Association, the numbers of those preferring the oven to the ground have skyrocketed, from a paltry three percent in 1960 to a whopping forty-three percent in 2012! It's an affordable, space-saving means of disposal without all the fuss and muss of a casket and hearse (urns fit comfortably in the trunk). Besides, I'd rather be perched on the mantle than stuck in the cold, hard ground any day.

THE NEED FOR ZZZZZZ
August 22

It happened again last night. I'm sleeping like a baby when all of a sudden, my brain stirs to life and shouts, *"Wake Up and Start Worrying!!"* It happens all the time. My internal alarm strikes 3 a.m. and that little worrywart inside my head jumps aboard the stress treadmill and starts ticking off the miles. Lately, he's been training for a marathon.

After recently resigning from my job and selling my house, I'm facing an impending move and a new direction in life. This provides a veritable smorgasbord of concerns on which my anal-retentive mind can feast. To make matters worse, I can never seem to sleep past dawn. In fact, I'm lucky to make it to 6 a.m. Usually I'm tossing and turning by four, checking and rechecking the clock, until boredom sets in, I throw back the covers, and get on with my day. Even when I aim to sleep late, my brain usually betrays me by whispering, *"Time's a' wasting!"*

All this insomnia got me thinking about the need for sleep. Sometimes I imagine how much I could get done if sleep weren't a necessity. I could write or work out or study till dawn. Unfortunately, my early rising usually means I'm comatose by ten. Sleep comes quickly, even if it's short lived.

So, let's take a peek at the necessity of sleep and remind ourselves of the critical role it plays in a healthy body. We are not alone in our need for Zs. Sleep is a necessity akin to eating and breathing, and all mammals share the same fundamental sleep patterns, which, in humans, are broken down into five stages, culminating in REM, or "rapid eye movement."

Stage one is the light sleep we experience as we drift off. This is also the time when sudden muscle contractions, known as *hypnic myoclonia,* can jolt us awake, since they are usually preceded by the sensation of falling. Stage two eases us into stages three and four, known as "deep sleep," during which our brain downshifts, producing the slow delta waves that accompany these stages. It's during deep sleep that it's most difficult to be awakened. And then comes the REM sleep.

Rapid eye movement sleep is the crème de la crème of snoozing and begins with signals sent from the base of the brain, or pons. The signals are whisked to the thalamus and then relayed to the cerebral cortex, that vital outer layer of the brain responsible for higher thought. As the cerebral cortex is stimulated, the pons sends other signals to neurons in our spinal cord, shutting them down so we don't act out in our sleep—a dangerous but, I'm guessing, potentially hilarious condition called REM Sleep Behavior Disorder. Can you imagine the carnage associated with an afflicted competitive eater??

REM sleep is when our dreams take flight. Each of us spends about two hours a night in the dream stage, which scientists have discovered is critical to a healthy brain. Protein production increases during REM sleep and those deprived of it have greater difficulty learning and retaining information. REM may also be fundamental to brain development since infants spend much of their downtime in this stage. (I wonder what fills their dreams. . . giant nipples, perhaps?)

In fact, sleep is crucial to good overall health. Sleep gives our cells time to repair and produce proteins and promotes a healthy nervous system. Sleep deprivation leads to lack of concentration, memory impairment, and, possibly, early death. Studies among rats show that well-rested rats typically live two to three years, while their sleep- deprived buddies usually croak after only five weeks.

Scientists are still trying to sort out why we dream. Some believe dreams are the brain's attempt to organize and interpret random signals given off during REM. Freud believed dreaming provided a safety valve for our unconscious desires and that through our dreams, we could fulfill our innermost wishes. However, they are constructed, dreams provide a momentary escape from reality. They can exhilarate, stimulate, or terrorize, depending on their content. But they can also provide a window to the past.

Sometimes, I dream of my parents. And in those dreams, they are young and smiling and alive.

Perhaps Freud was right.

BODIES IN SPACE
August 29

This week I write to you from the sunny coast of west Florida. I've been in Tampa for the past four days, taking a course in flight medicine and learning just how different (and dangerous) it is treating patients in a helicopter versus the earth-bound safety of an ambulance. There's a lot to learn.

The use of helicopters in emergency medicine is inherently dangerous, yet the benefits to the critical patient, which include an elevated level of care along with rapid transport to a trauma center, far outweigh the risks.

Aside from safety, there are a plethora of considerations when it comes to patient care in the skies, for we humans are acclimated to life on the ground. There are numerous ways our bodies are affected by elevation. Changes in altitude mean changes in pressure, which affect our body's ability to oxygenate itself. And the higher we go, the more complicated things become. This got me thinking about the effects of gravity (or the lack thereof). What happens to a body in space?

This question was at the forefront of space exploration. As the technology evolved to launch rockets into space, it was a natural progression to want to send humans. But what would happen to people in space? Would they survive? Would they explode? How would the body react to weightlessness? Nobody knew. So, as in all of science, if you can't experiment on a human, how about the next best thing?

On June 11, 1948, the first in a string of critter cosmonauts lifted off from White Sands, New Mexico. Albert I, a rhesus monkey, successfully survived his mission, as did his cohorts Albert II and IV (Albert III's fate remains classified). Unfortunately, the furry pioneers were rewarded with fiery deaths upon impact when they crash-landed back on earth.

The Alberts were the first in a long line of animal astronauts. They were followed by dogs, cats, mice, and spiders. And the Alberts were not the only propelled primates. Squirrel monkeys, Gordo and Goliath, were followed by a couple of famous chimps, the first of which blasted off on January 31, 1961. Ham became the first ape in space and his sixteen-minute flight was duplicated four months later by Alan Shepard. Both were later bested by Enos, the first chimp to actually orbit the earth. He was then followed by copycat John Glenn, who accomplished the task on February 20, 1962. Sadly, although Enos survived the rigors of spaceflight, he died eleven months later from (according to NASA) "non-space related" diarrhea. Sounds suspicious. . .

Humans have now mastered space travel and through exhaustive experimentation, we are recognizing myriad ways extreme altitude and weightlessness affect the human body. Let's start with my favorite, the bones.

Humans are bipedal creatures, but in the weightlessness of space, astronauts no longer walk, they simply float from point A to point B. The floating may be fun, but without the regular load bearing of walking, their bones quickly break down, releasing valuable calcium into the bloodstream. The bones become brittle, making them easier to break, and the excess calcium means greater risk of kidney stones—double trouble when you are stuck on the International Space Station two hundred miles above earth.

Muscles are another problem. That same load bearing that keeps bones strong also keeps muscles healthy, while weightlessness causes muscles to weaken and atrophy, which can lead to injuries. That's why astronauts make a point of exercising (no matter how silly it looks). Strapping themselves to a stationary bike or treadmill is critical in maintaining muscle mass, and scientists are constantly improving space-friendly nutritional supplements to complement the weightless workouts.

Aside from the bones and muscles, many of the body's systems are dramatically affected by space. Just like the other muscles in the body, the heart doesn't have to work as hard in weightlessness and, over time, it too can atrophy. Balance becomes another issue. The lack of gravity means the inner ear can't tell which way is up, leading to disorientation and motion sickness.

And what about sleep? Without the normal day-night cycle, the circadian rhythm of sleeping is thrown out of whack. Since you all read last week's post and are well versed in the importance of sleep, you can imagine the effect sleep deprivation has on our astronauts as they go about their daily chores.

Space is still a new frontier, despite our many missions. But our bodies are of this earth and have evolved as such. And although we may defy it, each system in our body relies on that fundamental force, gravity.

TAKING A BREATHER
September 5

Damn, life is complicated. As many of you know, I'm in the throes of a major life change. I've quit my job as an archaeologist, sold my beautiful little house in Cocoa, and next week I'll head to the lush hills of Tallahassee to hunt down a place to live. I'll also be teaching a course I've never taught at Florida State University, so I'm using this time off to prepare for my return to academia. Life was much simpler as an archaeologist—who knew unemployment could be so time-consuming?

With all these changes in the mix, writing a weekly post has become quite a challenge. When I was working as a public archaeologist, tucked in my cozy office in front of a computer, the blog was part of my daily routine and I would parcel off a bit of my afternoon for research and writing. My reward? Over forty-six thousand page views, to date. But now that I have no routine (as well as no office) and am pulled in numerous directions, it's become harder to devote the time needed for thoughtful, well-developed posts. So, I'm taking a breather.

After posting each week for the past year and a half (seventy-eight and counting!), I've found that the most recent posts are not always the most read. Thanks to key words and Google searches, I have found the back posts are just as popular as more recent additions. Therefore, I'm confident The Body Blog will continue to be read, even in the absence of weekly contributions. This would be a great time for you, the reader, to delve into the blog's archives, for we've covered a wide range of topics about the human body.

My background as a former firefighter-paramedic turned bioarchaeologist has provided a broad and unique perspective on the body and the myriad ways it is impacted by culture. We've explored such topics as capital punishment, body art, dental mutilation, and embalming. We've traveled back in time to explore the evolution of the bra, the condom, cremation, and the kiss. The body's systems have provided endless fodder and we've investigated just about every orifice and appendage, from vaginas and penises to the heart, belly, nose, and brain. We've covered the realm of sex: its evolution, selection, our development, and the convoluted history of syphilis. We've also delved into the darker side of anatomy, from quackery and grave robbing, to autopsies and necrophilia. Whatever you can do to or with a body, we've touched on over the last eighteen months. Even the ephemeral realms of empathy and exploitation, the shame of disfigurement, and the power of a mother's love.

So, as I take time to realign my life, I hope you'll peruse some of the older posts. I'll still add to the blog, just not at the frenetic pace of the past year and a half, and we'll continue to explore the fascinating entity that is the human body, together.

Thirteen Years. . .
September 11

The sun has lost its outline
As I wash the weary skies
The quarter moon ascends the slope
Where on its back it lies

Among the woods the cedar pines
Begin their sweet release
And buried in their needled brow
I've come to find my peace.

SIR ISAAC NEWTON, TRAUMA JUNKIE
September 26

Lately, I have been dreaming about trauma. In my spare time, I've been doing ride-alongs with Air Care, the EMS helicopter based out of Orlando's level 1 trauma center, just to get a taste of my former life as a paramedic. (Does it make me a bad person to wish for a critically injured patient?) You could say I've had trauma on the brain.

To me, trauma is the most exciting realm of emergency medicine. Traumatic injuries can result from a variety of scenarios, from crashes to falls to shootings and stabbings. But they all have one thing in common: physics.

We'll begin with a quick trip back in time. It's the mid-1600s and a less-than-stellar student named Isaac Newton is attending Cambridge University in England. We've all heard the story: he's reclining under an apple tree when suddenly, out of the blue, he is bonked on the head by an errant piece of fruit. The blow to his noggin results in his epiphany that gravity exists, that it can help explain the motion of planets, and that, if an object has enough mass, it can do serious damage to one's head.

OK, I made that last bit up. In fact, the whole story is fairly bogus. Yes, young Newton supposedly observed an apple fall to the ground (no mention of head injury), which may have triggered contemplation of the moon's orbit. However, it took him years of study, informed by the works of Descartes, Galileo and Kepler, to develop what would become some of the most important and impactful discoveries of motion and light.

So, what does any of this have to do with traumatic injury? A lot, it turns out. It wasn't so much Newton's theories on light and gravity that would influence trauma medicine some four hundred years later (although anyone who has taken a tumble knows the inherent drawbacks of gravity). It was his later work on motion, which he handily summarized in three laws, that would elucidate the drama of trauma and its devastating effects on the body. Since you are probably not interested in the mathematics behind Newton's laws, and the thought of explaining quantum mechanics is enough to send me screaming, we'll simply explore the basics of the three axioms and how they relate to the treatment of trauma patients.

Law number one: A body in motion remains in motion unless acted upon by an outside force. Say you are driving along one night when suddenly you veer off the road and strike a tree. In compliance with Newton's first law, if you were travelling fifty miles per hour, so were your internal organs. Your car strikes the tree, your body, if unrestrained, then strikes the steering wheel, and your internal organs slam against the confines of your body (brain against skull, heart against ribs—you get the picture). Each impact has consequences for the body's tissues. Therefore, seatbelts are of critical importance. Seatbelts provide a mechanism for keeping you from impacting the inside of the vehicle when it is brought to a sudden stop. Yes, seatbelts can do their own damage at high speeds, but I'd take a seatbelt injury over a steering wheel to the chest any day.

Law two: The force on an object is equal to its mass times its acceleration. Think about our collision—car vs. tree. The heavier the vehicle (more mass), and the faster it is moving (acceleration), the greater the impact when it strikes an object; especially if that object is also massive (say, an oak tree). Which leads us to the third law.

Law three: To every action, there is an equal and opposite reaction. Once again, in our collision we were travelling fifty miles per hour, which means that the tree we struck will push back with the equivalent force of fifty miles per hour of speed. Thus, the motto, "Speed kills." How much better off we would have been had we been putting along at a snail's pace. High speed impacts have devastating effects on the body. The breaking of bones, the shearing of vessels, and the subsequent bleeding that results from these injuries make trauma patients some of the most complex to manage, since any—and many—body systems can be involved.

Thanks to Newton, we can make predictions as to the types of injuries that will result from traumatic events. Whether it's an auto accident, a shooting, or a fall, if we know the speed of the vehicle, the caliber and velocity of the bullet, or the height of the fall, we can infer the potential damage inflicted upon the body. Thus, the "mechanism of injury" is one of the fundamentals of trauma assessment and it all originates with the genius of Newton.

Could Newton have imagined the impact his work would have on trauma medicine? Perhaps. The 17th century was a period of dramatic discoveries in science and medicine. The year 1628 saw the publication of Harvey's *An Anatomical Study of the Motion of the Heart and of the Blood in Animals*, which detailed his groundbreaking work on the cardiovascular system (although he scores no points for brevity of title). Frenchman Jean-Baptiste Denis and others were tinkering with blood transfusions and slowly refining their techniques—dog to human simply didn't work. And Leeuwenhoek would perfect the microscope, under which he would discover blood cells and microorganisms (along with the creepy crawlers inhabiting his own dental plaque). In this climate, it would seem natural for Newton to appreciate the impact (no pun intended) of his laws on medicine. And if he actually was hit on the head with an apple, surely, he could envision the carnage, had that apple instead been a cannonball.

A BRAINY NEW YEAR!
January 1

Now that 2014 has drawn to a close and we look ahead to the coming year, it is time to partake in that tried-and-true tradition of all wishful thinkers, the New Year's Resolution. Let me guess: you're intent on losing those extra pounds (even though you probably started accumulating them back in 2010); you're going to dig out those musty workout clothes and force yourself to the gym (if you could only remember how to get there); and you swear to give up at least one of your vices, whether it be cigarettes, junk food, or, as in my case, gin.

The New Year is a time for reflection. We look back over the past twelve months, at the changes in our lives (or lack thereof), and the swiftness with which each year passes, anticipating what lies ahead as the year unfolds before us. It can be a scary time.

Let's face it, the older we get, the shorter our future becomes. The horizon, which seemed so far off as we cruised through adolescence, suddenly looms large before us as we find ourselves cresting the onrushing midlife wave. And with each passing December, with each approach of the New Year, we grow more aware of the "tick-tock" of Father Time.

I happen to love the New Year. To me, it signifies a fresh start, new beginnings, and the opportunity to set new goals. I know that spring is just around the corner, putting an end to the winter doldrums which, here in Florida, last for an excruciating week and a half. But all this anticipation got me thinking about that one crucial commodity for planning ahead: optimism.

Optimism is a vital mental tool, one that can not only make us feel better, but can actually improve our health. Some evolutionists believe the tendency toward optimism—what they term the "optimism bias"—is hard-wired into our brains and was integral to the dramatic cultural transformations that have taken place among humans over the past fifty thousand years. In that short span of time (geologically speaking), we have gone from artless, illiterate hunter-gatherers to beings that communicate on a global scale, create masterful works of art, and traverse the cosmos. And how could we have achieved any of this without a strong dose of optimism?

The optimism bias is universal among humans but can vary depending on our individual wiring. Even though our brains are constructed as two mirrored halves (bilaterally symmetrical), many of our skills, such as language and handedness, are controlled by a designated half. This brainy asymmetry allows us to perform many tasks at once. And when it comes to optimism, it is the left hemisphere—or "left brain" in neuro slang—that takes the lead.

This "lateralization" of optimistic behavior is expressed in the way we think, feel, behave, and even plan for future events. We are all familiar with the "glass half full-half empty" analogy. Optimistic folks tend to focus on the positive, usually ignoring or minimizing anything that threatens to quash their rosy outlook. Pessimists (those "right brainers") do the opposite: their "half-empty" mentality tends to lead them down the gloomy paths of worry and doubt. These opposing life views have even been tested experimentally. Optimists will spend less time focusing on negative visual stimuli than their pessimistic counterparts, who not only spend more time focusing on the negative, but also tend to take greater cues from negative stimuli in their environment.

That's not to say that pessimism should be eliminated. It serves a vital role in keeping our overly optimistic tendencies in check. Can you imagine the havoc that would ensue if we failed to anticipate setbacks, accidents, or illness? Unbridled optimism could result in financial hardship, traumatic injuries, or debilitating illness if we ignored the necessities of savings accounts, seatbelts, and preventive medicine. A little bit of pessimism goes a long way.

So, as you enter the coming months and tackle your resolutions, strike a balance between your brainy halves. Approach the New Year with the utmost optimism but temper it with a small dose of pessimism. Perhaps you will finally lose that weight, or become a model of fitness, or finally shake that vice. But if you don't succeed, at least take comfort in the fact that you are making an effort. And remember, the key to optimism is that vital, life-sustaining force: hope.

Happy New Year!

THE MYTHOLOGY OF MASTURBATION
January 8

Here's an embarrassing question: When was the last time you masturbated? According to statistics, most of my male readers have indulged within the past thirty days (and if you are a teen, I'm betting within the last thirty minutes). As for the females, the percentage is smaller, but still consistent. Let's face it, we've all done it, yet no one ever wants to admit it. Why is that?

For some reason, masturbation cannot seem to escape the bounds of stigma, perhaps because it's been associated with so many ridiculous fallacies. So, in an effort to bring masturbation out of the closet, let's put some of these ludicrous misconceptions to rest. Let's begin with terminology.

The slang we've devised to avoid uttering that embarrassing word, "masturbation," has taken on a life of its own (kind of like the plethora of names you boys have concocted to sidestep the word "penis"). Whether you're "spanking the monkey," "jerkin' the gherkin," "cuffing the carrot," or, my all-time favorite, "buffing the banana," it goes without saying that you're participating in that tried-and-true pastime, "solo sex." And the fact that the majority of slang is masculine indicates you fellas spend far more time dwelling on the subject than us girls. And then there are the myths. . .

The most common myths about masturbation associate it with some form of physical disorder. Blindness, hairy palms, impotence, or a curving penis are just a few of the malignancies that have been linked to masturbation. But I'm here to assure you, this is complete nonsense. No, you will not go blind, your palms will remain hairless, your impotence will originate from some other cause—be it age, medication, or illness—and if you sport a curving penis, you were probably born that way.

Side note: a crooked tool is just as effective as a straight one, so don't sweat it if yours subtly points east or west.

The myths not only affect the boys. Girls have been warned about infertility, a loss of their virginity, and an inability to climax during partner sex, should they choose to fly solo. But like the male myths, these too are pure rubbish. In fact, sex therapists claim that masturbation may make it easier to climax during partner sex—for girls and guys. Think of self-stimulation as a proving ground: the more you practice, the better you know your own body and what it takes to bring you to that pinnacle of intercourse: orgasm.

"But what if I have a fully functioning partner?" you ask. Masturbation should not be considered a form of cheating. Solo sex falls into a different realm than partner sex. It's a means of indulging yourself, whether for relaxation, stress relief, or when you simply need a little "me" time. Granted, you and your partner must come to an understanding, lest someone's feelings get hurt. But part of being a good partner is giving each other space, and masturbation is a safe and easy means of satisfying oneself, especially on those occasions when your significant other lacks the time or energy to see to your needs.

One of the biggest hurdles to masturbation is the guilt that some feel when indulging. Whether this stems from your religious beliefs, your parents, or the simple embarrassment associated with certain bodily functions, rest assured that masturbation is part of a normal, healthy sex life. And the benefits abound.

Aside from fine-tuning our ability to orgasm, masturbation reduces stress, alleviates sexual tension, improves sleep, and is a fine alternative for those lacking a partner. And as long as it doesn't replace partner sex or become an impediment to daily life, it's the safest sex around!

So, the next time you feel the urge, indulge yourself. Shuck off the guilt and spend a little quality time with your body and I assure you, you'll sleep like a baby.

IS BARE BETTER?
January 23

This morning, for some strange reason, I woke up thinking about shoes. I wandered into my closet, curious about the number I had accumulated over the years and was astonished to find thirty-two pairs. How the hell did so many shoes get into my closet?!

Americans love their shoes, as do most societies where fashion trumps practicality. Do we really need so many shoes? Women are the worst culprits. It's a luxury to have a vast selection to choose from, and fashion forbids us from pairing the wrong shoes with our outfits. Think of the anarchy that would ensue should we pick strappy sandals over pumps. Oh, the humanity!!

But shoes didn't start out as tools of fashion. The shoe evolved out of sheer necessity. Try to imagine venturing cross-country with nothing between your feet and the torturous ground on which you tread. Rocky outcrops, razor-like switch grass, not to mention the temperature extremes of blazing deserts and icy tundra, would be pure hell. I'm guessing our clever ancestors took to shodding their feet soon after they discovered the benefits of clothing. If you can protect your torso by wrapping it in hides, why not your feet? They were no dummies, our ancestors.

One of the oldest shoes was found a few years ago, tucked inside an Armenian cave. Dated to over fifty-five hundred years ago, this leather moccasin was made from cowhide and laced with a leather cord. It had the good fortune of being buried beneath a dense layer of sheep dung, thus the exceptional preservation, but footwear goes back much further in human history.

One of the earliest forms of shoes appears to be the sandal, which is not surprising, considering its basic design. Ten-thousand-year-old sandals have been recovered from Fort Rock Cave in central Oregon, proving again what excellent preservation caves afford.

Curious sidenote: Do we find these ancient shoes in caves because people happened to be living there, or were caves the precursors to today's walk-in closets? Something to think about.

But back to the Fort Rock sandals. They ranged in size from adult to child and showed the wear and tear of countless miles, some still sporting the mud from prehistoric landscapes. Some even held the imprint of its owner's foot: faint impressions of a hunter who traipsed through the Oregon woods in search of game.

And speaking of feet, some of our evidence for shoes comes directly from studying the feet of our ancestors. Scientists have noted changes in the bones of our feet that may have occurred secondary to adopting footwear, possibly going back forty thousand years. Even today, shoes are reshaping our feet. Unfortunately, many of these changes are detrimental. In fact, for all the protection shoes can afford, many of the styles we sport come with repercussions.

Some evolutionists believe shoes have made our feet less healthy. The excellent support afforded by cushioned soles may feel luxurious but can actually lead to weakened arches and flat feet. Thick soles also cause diminished sensation between foot and ground, which means lack of communication between foot and brain, putting you at risk for missteps and stumbles. And don't even get me started on high heels. Yes, those stilettos may look sexy, but they're doing your body no favors. Stress fractures, bunions, and hammertoes are just the beginning of chronic problems associated with high heels, not to mention back strain and deformation of your calf muscles, which become shortened with excessive wear.

While high heels present problems, so do their counterparts, the flats. Ballet flats, which are all the rage, provide little to no shock absorption, a real problem especially for those flat-footed individuals. They can also cause tendonitis and heel pain and provide scant protection against bottle tops and broken glass. But one of the biggest culprits is the flip-flop. Here in my home state of Florida, flip-flops are as prevalent as retirees (and frequently found in unison). The flip-flop is the next best thing to being barefooted, which is great for those sunbaked beaches, but provides little protection in the real world. Most are too thin and leave the foot exposed to environmental hazards—such as that friend who chooses the stilettos—and, let's face it, they are a far cry from attractive.

So, the next time you reach for a pair of shoes, choose wisely. Select your shoes as you would select a mate: ones that are well made, practical but stylish, and most important, the perfect fit.

THUMBS UP!
February 6

Here's an experiment: try going fifteen minutes without using your thumbs. Bet you can't do it. We may not give them a second thought, but life would be very difficult without thumbs.

Last week, my left thumb was knocked out of commission after I sustained a painful bite from one of my lovebirds (they really deserve a more appropriate name... devil birds, perhaps). I introduced you to my birds, Tuukka and André, in last February's *Bird Brains*. Well, a few days ago, in an attempt to prevent their close encounter with the ceiling fan, I was trimming their wings. I had Tuukka wrapped in a towel and flipped on his back, but even with the protection of the towel, he managed to swivel his evil little head and take a hunk out of my thumb. To make matters worse, he tore off a small chunk of skin that for the next few days caught on everything it came in contact with. Putting on socks, toweling off from the shower, getting ice from the freezer—every simple task became a painful ordeal as I tried to manipulate my maimed thumb. It got me thinking about our dependency on thumbs and the crucial role they have played in our evolution.

First, we must start with the hand in general. Although we now walk on two limbs, we evolved from four-legged stock, thus we are tetrapods at heart. And the majority of tetrapods sport limbs with five (or fewer) digits. Yes, many a tetrapod has lost a digit here or there. Bat fingers still come in fives but are draped in a leathery wing. Bird fingers come in a bizarre array of digits, depending on the species. And horses' feet have been whittled down to a single lonely toe. But we humans have maintained the standard five and boast an especially talented member, the fully opposable thumb.

Although our hands and those of our closest relatives, the chimps, are similar in structure, our hands—especially our thumbs—have several key advantages. Our thumbs are longer, stronger, and more maneuverable than those of our primate cousins, whose thumbs lack the musculature of our mighty first digits. And because they are so short, the chimps' puny thumbs cannot "oppose" their other fingers, and we all know how vital opposable thumbs are. That's why you'll never catch a chimp "in a pinch." They simply lack the ability. Instead, they are forced to press their miniscule thumbs against the sides of their index fingers when picking up small twigs or snatching bugs from the forest floor.

Our opposable thumbs allow us to grip; a crucial skill in making and utilizing tools. Scientists claim the evolution of our "power grip" was crucial to wielding clubs and throwing stones; tasks that would have come in handy when warding off predators and taking down prey. The evolution of the "precision grip," also made possible by a strong, flexible thumb, would have enhanced toolmaking, allowing our ancestors to construct the intricate objects that would accompany the rise of *Homo sapiens*.

Powering our thrifty thumbs are three muscles lacking in the chimp hand. These muscles provide strength and control, and the small saddle joint on which our thumbs sit is key to its full opposability. So, unlike chimps, we can pinch, pluck, and snap to our hearts' content. In fact, scientists are now examining the role of thumbs in human evolution and, it turns out, those with stronger, more agile thumbs may have had an evolutionary edge over their weaker-thumbed cohorts. The ability to produce more effective tools in greater numbers may have edged out the competition among our hominin ancestors, which makes sense. Gals usually go for the bigger, better tools.

But what about in today's modern society, where tool production has fallen by the wayside? Fewer folks actually produce their own tools, much less make their own clothing, build their own houses, or grow their own food. Do our thumbs still present an evolutionary advantage?

Perhaps if you are a habitual gamer. Just ask any gamer the importance of thumbs and I'm sure they'll present a litany of benefits their nimble thumbs afford. Whether they are blasting their way through Doom, leading expeditions across Monkey Island, or taking down foes in Mortal Kombat, the faster the thumb, the better their chances of conquering the universe.

Unfortunately, this nerdy set of skills probably doesn't confer much of an evolutionary advantage. If history is any indication, I doubt the gamers will be outbreeding the rest of us anytime soon.

PICTURING THE DEAD
February 20

Think back to the funerals you've attended, particularly the open casket affairs. Do you remember details of the deceased? Their expression, the tone of their skin, the contours of their face? One final question: did you take a picture?

In today's photomaniacal culture, we take pictures of everything: our pets, our food, and, most often, our body parts. But we still tend to shy away from taking pictures of our dead. I find this curious, for it wasn't always so.

For thousands of years, burials—at least of elites—have been accompanied by some sort of remembrance. Long before photography, this came in the form of death masks. Made from an array of materials, wax, plaster, or in some cases, precious metals, these masks memorialized the dead as they appeared in life.

One of the oldest and most famous was that of King Tutankhamen. The young king, who died around 1400 B.C., was interred with a splendid gold mask weighing a whopping twenty-four pounds, although the accuracy of the mask is debatable. Since Tut's time, death masks have been a common means of memorializing the dead. From Mary, Queen of Scots in the 1500s to Sir Isaac Newton in the 1700s and, closer to home, Civil War hero and president, Ulysses S. Grant in the 1800s, these masks capture intricate details of the deceased, as they appeared prior to interment.

But all this changed when photography burst on the scene. In 1837, Frenchman Louis Daguerre, using a concoction of silver-plated copper, silver iodide, and mercury, was able to create the first permanent image. Referred to as Daguerreotype, this early photographic technique, like most new trends, began as an expensive luxury. Louis' early photography captured images of family and friends and was more expedient and practical than hiring a portrait artist. Parents finally had a means of recording their broods for posterity. The only problem was that many broods did not survive to adulthood. Combine newly developed photo ops with high child mortality and you have a recipe for a morbid yet practical new Victorian fad, postmortem photos.

Known in Latin as *memento mori*, these ghoulish keepsakes became fashionable on both sides of the pond. At a time when family photos were a desirable commodity, capturing images of the deceased, especially those of children, took on a whole new meaning. The death of a child meant little time to record the child's image, so when death came swiftly, so did the postmortem photographer. The child was traditionally posed in lifelike manner, sometimes alongside favorite toys, or cascades of flowers, or tucked among living family members. Over time, as new techniques made photographs easier and less expensive, even those of modest means could capture images of their dearly departed.

Postmortem photos also accompanied the taming of America's Wild West. What better way to publicize the death of infamous outlaws than to exhibit photos of their corpses? From the Dalton Gang (whose bank robbing was cut short when the docile townsfolk of Coffeyville, Kansas, gunned them down), to Jesse James (same business, same bloody end), to George "Bittercreek" Newcomb (former member of the Dalton Gang who was offed by his amigos in exchange for a fat bounty), photos of dead outlaws served as proof of the inherent dangers of criminality on the frontier.

In today's modern culture, postmortem photos fall into a singular category accepted, even coveted, by a morbid public: the dead celebrity. We seem to crave photos of the famous who have met their ends. There are even websites dedicated solely to dead celebs. Celebritymorgue.com sports photos of an array of famous corpses, from loveable dictators like Mussolini and Stalin to ghastly morgue shots of Marilyn and Tupac. Even nonhuman notables have made the cut. P.T. Barnum's prize-winning elephant, Jumbo, was memorialized in a postmortem photograph after he was inadvertently struck by a train while moseying across the tracks (poor Jumbo).

Postmortem photographs have evolved through the ages. What was once a respectable, albeit macabre, means of memorializing the dead has morphed into a tool for gawkers and sensationalists. But even in its current twisted state, postmortem photography has a way of satisfying a visceral desire in all of us: the chance to look death in the face.

THE HUMAN TOUCH
February 27

God, I love hockey. As you may have read in November's *Size Matters*, I'm a devoted hockey fan. (Go Lightning!) I love the speed, the strength, and the beautiful biomechanics of this amazing sport. I also love the rituals following each goal. Here's how they unfold: the puck hits the back of the net, the shooter raises his stick in triumph, and the team then consummates their achievement with a massive group hug. Fellow teammates pile on in what can only be compared to an anaconda mating ball. It's a raucous love fest, which is all the more ironic considering the brutal physicality of the sport.

The celebration doesn't end there. Once the bro-bundle disassembles, the shooter and his line then glide over to tap paws with their benched teammates. But the pinnacle of all hockey celebrations comes at the end of the game when the winning team congregates for what I call "the goalie kiss." The victors line up in front of their goalie and one-by-one bump helmets, a means of acknowledging his skill. It's the closest thing to a man-kiss you'll ever see in American sports (or America, for that matter).

Hockey is not the only sport that involves teammate touching. In fact, most team sports involve some degree of touch, whether to celebrate an achievement, allay a mistake, or simply encourage one another. A team's touchability has even been linked to their ranking: the more successful the team, the more they tend to touch (although it's the chicken-or-the-egg conundrum as to which comes first). And why do they do this? Because touch is an integral part of human communication.

Our sense of touch begins in our skin. Nerve endings that originate in our dermis send messages via our spinal cord to our brain, where the information is processed. Our brain has evolved two separate pathways to analyze touch. The primary somatosensory cortex deciphers the fundamentals of touch: pressure, texture, vibration, and location, which is critical for navigating our world. But the second pathway is just as critical, for without it, we would respond to external stimuli like automatons. I'm talking about the emotional aspect of touch.

Our complex brain performs a remarkable sensory feat each time we engage in touch: it places that touch in context. It does this by utilizing particular sensors in the skin, which trigger regions of the brain associated with pleasure, pain, and social bonding. And it's the brain's dual pathway that explains why the same type of touch can be perceived in widely disparate ways.

Imagine the touch of a loved one: the reassuring pat of a parent, the warm hand of a child, or the sensual stroke of a lover. These contexts engage sensory fibers that trigger emotional bonding reflexes within our brains. Now compare that to the eerie touch of a drunken stranger who sidles up to you at a bar. Same touch, very different scenario. Our reactions to touch are based on the emotional interpretations produced in our brain. And when it comes to touch, it's all about context.

Touch is more than a means of engaging our world, it is fundamental to our emotional development. Children deprived of touch not only suffer emotionally, but its lack affects their immune response, digestive health, and their ability to integrate in society. That's because touch forges trust: a response rooted in the chemicals within our brains. In the proper context, touch triggers the release of oxytocin, a hormone closely associated with our sense of trust, which explains its role in sex, birth, and breastfeeding. A warm touch also reduces stress by tamping down one of the key stress hormones in the body, cortisol.

And it's these positive benefits that drive much of the recent research on touch. At the Touch Research Institute (yes, there is such a place) at the University of Miami, scientists are hard at work exploring the emotional benefits of touch. According to their research, touch, in the form of therapeutic massage, can alleviate headaches and anxiety, help with muscular and spinal cord injuries, and reduce stress and pain. The Institute is even exploring how regular massage can ease postpartum depression. It turns out that massaging an expectant mother reduces stress and depression during and after pregnancy, but also benefits the baby by lowering the incidence of premature births and low birth weight among tots. So, if your significant other is expecting, be a dear and give her a rub.

They say a picture is worth a thousand words. The same can be said of touch, only the language of touch goes beyond mere words. Touch speaks to us on a visceral level, stirring emotions that drive us as human beings. Trust, desire, security, and well-being can be relayed without uttering a sound. All it takes is the right touch.

The Body Blog

WELCOME TO THE GUN SHOW
March 6

There's nothing more striking than well-defined muscle. The firm bulge of biceps, the ropey thickness of quads, the ravishing ripples of six-pack abs. Few things compare to the beauty of lean muscle. In fact, I'm willing to overlook certain character flaws (kleptomania, bizarre fetishes, or—god, forbid—an aversion to hockey) in exchange for a ripped physique.

Not only is muscle beautiful, it's also delicious. You may not give it much thought, but each time you bite into a juicy burger, feast on a platter of wings, or tear into a pile of chops, you are ingesting the muscle of some critter, be it beef, bird, or swine. Let's face it, muscle rocks. So, to pay homage to the magnificence of muscle, let's explore these wondrous tissues and the many roles they play in our bodies.

Were we to inventory the six-hundred-plus muscles that make up the human body, it would take the better part of the day. That's because muscles come in an array of forms and sport tongue-twisting names based on their characteristics. Some are named for their size, such as the largest in the body, the gluteus maximus, which you are probably sitting on right now. Some are named for their shape, like the deltoid, because of its triangular silhouette. And some derive their name from the direction in which they run, like the beautiful rectus abdominus that extends vertically along the belly, forming those lovely little cans within the six pack.

The body contains three different types of muscle. Skeletal muscle is what gives the body its beautiful design. These voluntary, striated muscles move our bodies by manipulating our skeletons. By pulling on bone, skeletal muscle enables us to walk, run, blink, and smile, swivel our heads, rotate our arms, and contort our bodies in myriad ways. Every movement is made possible through the contraction of these amazing fibers. And the anchor points for many of these muscles sculpt our skeletons, for wherever you have muscle pulling on bone, you have a bony prominence on which the muscles gain purchase. The larger the muscle, the larger the attachment site. So, as you work out, you're not only building muscle, but you're also building bone, as well.

Just as skeletal muscles move our skeletons, smooth muscle, otherwise known as visceral muscle, also plays a role in movement, but on a much finer scale. The blood coursing through our vessels, the food moving through our digestive tract, the air entering our lungs—each movement is dictated by our brain, coordinated through our nervous system, and carried out involuntarily via these rarely contemplated, seldom-seen muscles. Smooth muscles lack striations and are relatively weak. But although skeletal muscles get all the attention, smooth muscles are the true "movers and shakers" of our body systems. They deserve a bit more respect.

The third type of muscle drives the core of our being: the heart. Cardiac muscle is a bit of a hybrid. It shares some similarities with skeletal muscle, some with smooth. It has striations (like skeletal muscle) and is controlled involuntarily (like smooth), but cardiac muscle can do something no other muscles in the body can do: it can generate a pulse.

Specialized cells within the heart generate electricity, causing the cardiac muscle to contract. Cardiac muscle cells are arranged so that they overlap, forming a continuous web through which the electrochemical signals can pass. This causes the muscle to contract in a wave, drawing blood in, pushing blood out. And it's this beautifully synchronous motion that produces the apex of all life sounds: the heartbeat. Your heart will beat on average one hundred thousand times per day, thirty-five million times per year, and more than two and a half billion times during your lifetime (depending on your longevity, of course). That's a lot of pumping, which explains why a heart can wear out and why it's so important to keep it healthy.

The muscles in your body make up about half your overall weight. And because they are denser than fat, a fit individual can outweigh his unfit counterpart (and look a whole lot better doing it). And for those of you who have recently fallen out of the habit of working out, take heart: it takes twice as long to lose muscle as it does to gain it, so get up and get lifting! Muscles are fast learners with great memories.

Although our skeletons form the scaffolding on which our body systems are built, it's the muscles that bring our skeletons to life. Muscles move us, sustain us, and enable us to express ourselves in numerous ways, from simple gestures (a touch, a wink, a smile) to wondrous physical feats (walking, running, and lifting).

So, treat your muscles as you do your favorite pet: nurture them, nourish them, and give them plenty of exercise, and they'll repay you with a lifetime of unconditional love.

THE CLIMB OF YOUR LIFE
March 13

This morning on the radio, I heard one of my favorite songs, Edwin McCain's *I'll Be*, which has a great line in it: "I'll be better when I'm older." It got me thinking about age and perspective. It seems throughout early life, all we wish for is to be older. It's as if we are climbing a ladder and life will truly begin once we reach that next rung. So, I want you to perform a thought experiment: climb down the ladder and go back in time. Try to recapture your perspective as you ventured forth on your ascent.

When we are small, all we want is to be big. And the way to get bigger is to accumulate birthdays. That's why kids will never respond with "four" or "seven" when asked their age. They are "four and a half" or "seven and three-quarters!" Ask any kid and I bet they tack on that imperative fraction. But childhood is not a time to rush. Important things are happening in those little bodies. Although our growth rates are no match for the rapid development of infancy, we will still chalk up about two inches per year until we hit adolescence. Aside from growing, our bones are fusing, our teeth are erupting, and our brains are making critical connections that will help us read, write, and express ourselves throughout our lifetimes.

Then comes the day our birthday cakes boast double-digit candles. It's a magical time, adolescence. Hormones are raging, new hair is sprouting, and suddenly our bodies possess strange and wonderful abilities (especially if you sport a penis). And how do we respond to these mystical metamorphisms? By wanting to be a grownup, so we can take them out for a test drive. As teens we crave independence, the chance to make our own decisions, to be taken seriously as adults. We long to be part of adult society: by voting, serving in the military, or buying beer. As for our bodies, growth is winding down, the last of our molars are settling in (or being yanked by a dentist), and our reproductive capacities are given their final tweaks in preparation for parenthood.

When we finally make it to our twenties, a strange thing happens. Suddenly, the climb accelerates. Those rungs on the ladder go slipping by, greased by some unseen hand. You barely enjoy the freedom of maturity before thirty rears its ugly head. You are shocked to find yourself a parent and can't quite remember how you got here. You are saddled with a job, a spouse, and a mortgage, and before you know it, *Hello, forty!*

Forty arrives and you take a look around from your lofty perch and can't believe how high you've climbed. The air is cooler, it's easier to breathe, now that you've gained some perspective, and many of those imperative life decisions are behind you. Think of all you've learned! You look back with wonder at the antics of your youth: the foolish stunts you pulled, the poor judgment you exercised. It's a wonder you made it this far. And just as you are settling into this comfy locale, fifty arrives and practically knocks you from your rungs.

How can I be half a century old? you ask yourself. *Impossible! Why, just yesterday, I was graduating from high school. How could this much time elapse without my noticing?* And as for that view from the ladder—we're talkin' nosebleeds! The horizon stretches before you in a hazy blur, the objects on the ground, miniscule. You think back to your previous ideas of fifty and realize you were wrong all along. *Fifty's not old!* you tell yourself. *Sixty, maybe, or seventy, if I'm lucky to make it that far. Besides, if I live to be one hundred, I'm only halfway there!* Take a deep breath. . .

As someone who recently bid farewell to her forties, I cannot lend further perspective on the ladder of life. I'm still climbing, careful as I go. The best advice I can give is to enjoy each and every rung. Yes, you'll be challenged along the way, by love, loss, and hardship. But the higher you go, the luckier you are. Although the rungs of our youth grow small beneath our feet, always remember: the view from the top is mighty fine.

I'll be your crying shoulder
I'll be love's suicide
I'll be better when I'm older
I'll be the greatest fan of your life...
                    Edwin McCain

CARNIVORE KIN
March 20

Is there anything more annoying than a vegan? Since when did eating a chunk of cheese spell the demise of civilization? Don't get me wrong, I stand with the v's on their animal rights platform and abhor the trend in industrial farming. In fact, my precious niece and her lovely wife are vegans and, although I admire their integrity and applaud their dedicated activism, I have a distinct urge to stuff an egg down their throats. I just don't get the vegans.

As a committed carnivore, I must stand up for the meat eaters. We humans evolved to eat meat. Don't believe me? Reach over and pry open the jaws of the person next to you. What you'll see is living proof that we are meant to consume a wide range of foods—many of which had parents. The proof is in our teeth.

You can tell a lot about a critter from its teeth. From paleontology to paleoanthropology, teeth afford a handy way of identifying fossils, be they archaeopteryx or australopithecines, for teeth say a lot about how an organism lived and, most importantly, what they ate.

Close your eyes and picture a crocodile, preferably with its mouth open. What you should see in your mind's eye are teeth that vary in size but are all the same shape. The homodontous dentition of a croc is designed for one thing: grabbing flesh, which they do with lethal precision. Crocs don't chew; thus, they have no need for molars. They simply grab hold of that wildebeest, crush, and tear what they can, then swallow as big a chunk as possible.

Now picture your own teeth. They come in an array of shapes and sizes, for they have various functions. Our incisors and canines are for biting and tearing, our molars, for chewing. As omnivores, our heterodontous array opens us up to myriad foodstuffs, from seeds and nuts to plants and the all-important meat. And when I say meat is important, I don't just mean for today's burger munchers. The consumption of meat played a critical role in human evolution.

Some of the earliest evidence for meat eating comes from the dusty plains of Gona, Ethiopia, where butchered animal bones dating back to over two and a half million years ago were discovered in 1994. Carnivory was pushed back another million years by a discovery a few years back in nearby Dikika, where a goat-like critter was butchered almost three and a half million years ago. And what benefit would meat eating incur? None, really, unless you are interested in evolving a giant brain.

Meat eating meant a valuable source of protein for our hominin ancestors, which is critical for certain bodily functions, especially a metabolically demanding brain. Our brains consume about twenty percent of our overall energy intake. By exploiting high-quality foods like *meat*, our ancestors were able to supply their expensive brains without spending the majority of their day grazing like gorillas.

Our gorilla cousins are forced to spend endless hours munching their way through the forest in order to obtain the nutrients required for survival. And to process all that vegetation requires an enormous gut, thus their Buddha-like physiques. For the hominins, better foodstuffs meant a reduction in our guts. And since all of evolution is a tradeoff, smaller guts requiring less energy may have freed up fuel to grow our bigger brains. Not only did our guts get smaller, so did our teeth. As our teeth got smaller, so did our faces; thus, we lack the forward-jutting snouts of our ancestors.

More meat meant more people. A higher-quality diet, combined with the benefits of cooking (which boosts nutrients and kills pesky pathogens), would have enabled earlier weaning of infants, allowing women to have more babies more often, thereby spreading humans far and wide.

Meat also fueled our social evolution. Communal hunting instilled cooperative behavior and communication. As populations got bigger, meat on the hoof would have become scarcer, thus the domestication of animals provided a steady supply without the need to hunt. A steady supply meant populations could grow even larger, ushering in complex society, social stratification, and, eventually, industrialization; all made possible because of our love of meat.

So, the next time you are enjoying a juicy steak and some disapproving vegan gives you the stink-eye, take heart. If it weren't for our carnivorous ancestors, we'd still be wandering the African plains with our dinky brains and a fistful of tubers.

Hurray for the carnivores!

FIRE IT UP!
March 27

In the mood for a revelation? Ask firefighters how they feel about fire. Their response may surprise you. Firefighters spend their entire careers laying their lives on the line. Whether they are battling ripping house fires full of toxic combustibles, hiking mile after treacherous mile to combat raging wildfires, or sacrificing it all amidst the horror of a terrorist attack, firefighters are on the front line when it comes to battling the lethal force that is fire. So, you might find it curious that firefighters actually love fire.

Are they crazy? (Yes.) Are they obsessed? (Most definitely.) Or are they simply adrenaline junkies? (C. All of the above.) Firefighters are an unusual breed. Think about it—cops aren't infatuated with the criminals they cuff. Oncologists aren't enamored of the cancer they annihilate. So how can firefighters love what they spend their whole lives fighting? The answer: because they are human. And there's something about fire that humans simply adore.

As a firefighter-turned-archaeologist, I've spent a lifetime preoccupied with fire. As a firefighter, I saw its lethal side. I waded through charred wreckage, broke the news to grieving loved ones, and saluted the caskets of fallen comrades. As an archaeologist, I've explored fire's positive dimensions: its deep human history, the fundamental role it has played in culture, and the visceral connection we have with this phenomenal force.

Perhaps it's our ancient association with fire that has so ingrained it in our psyche. Evidence for its use goes back almost a million years, far longer that our species has roamed the earth. *Homo erectus* appears to have been the first to habitually use fire, and we find their ancient campsites, replete with butchered bones and beautiful stone tools, dotting their primordial landscapes. Fire provided warmth for our ancestors, despite the frozen grip of repeated ice ages, and gave them protection against predators stalking their primitive campsites. Fire provided light during their primal nights and formed the nucleus for social gatherings, where they exchanged information, manufactured tools, created art, and told stories. Fire was a catalyst of human culture.

As modern humans arose in Africa some two hundred thousand years ago, venturing forth to lay claim to the globe, they adopted fire, making it one of the most essential tools in their prehistoric toolkit. But the most important application for fire predated the arrival of *Homo sapiens*. In fact, we moderns may have never evolved had it not been for the invention of fire's most essential role: cooking.

Cooking transformed us. The advent of cooking, especially of meat, was pivotal in the evolution of our species. Meat provided the valuable nutrients necessary to fuel our ever-expanding brains, but it was the cooking of meat, along with plants, tubers, and anything else our ancient brethren happened to toss on the barbie, that streamlined our digestive tracts and fueled our giant brains.

Cooking transformed humans because cooking transforms food. If you don't believe me, hack off a hunk of raw sirloin and give it a chew. When you're finally able to swallow (about twenty minutes from now), I'm betting you'll be requesting the rest of that steak "medium well."

Cooking jump-starts digestion. In meat, it does this by breaking down the muscle fibers. Cooking also makes meat safer. The heat of cooking kills off pathogens, such as *Clostridium* and *Staphylococcus*, which hide out in undercooked meats just waiting for the chance to sabotage your gut.

Cooking also releases nutrients and calories—not only in meat, but in vegetables, as well. And aside from the nutritional benefits of cooking, most foods simply taste a whole lot better when cooked.

Which would you prefer? A raw potato eaten apple-style, or a steaming baked spud covered in butter and sour cream? (OK, I'd eat my shoe if it were covered in butter and sour cream, but you get my point.)

Cooking, like fire, drew people together. Even today, humans love to congregate and cook. What's more fun than hovering around the grill, surrounded by the silky fragrance of cooked meat and wood smoke, or assembling in the kitchen on Thanksgiving, as that golden-brown turkey emerges from the oven? Cooking brings us together, forges social bonds, and encourages sharing—all traits necessary for human society.

As for firefighters, the opportunity to sit down at the end of a busy day and swap stories over a hearty dinner plays a fundamental role in the cohesion and morale of a fire station. And ever since I traded my helmet for a trowel, dinner just hasn't been the same.

OUR SYMBOLIC SKULLS
April 3

If you had to name your favorite bone, which would it be? (Perverts, keep your responses to yourself.) I'm betting most of you would name the skull. Let's face it, when we think of bones, the skull naturally comes to mind (despite the fact that it's actually twenty-plus bones). There's not much regard for the limb bones, although I'm personally enamored of the femur. The patella gets little attention, probably because of its close resemblance to horse dung. And you hardly even notice your little coccyx, unless you happen to fall on your butt (who knew such pain could arise from such a tiny clump of bones?).

The skull is iconic. Not only for signifying the skeleton in general, but as a symbol that has infiltrated our culture. So, I thought it would be fun to explore how skulls crop up in everyday life and the quirky history of this popular emblem.

Long before the mass production of art, our ancient brethren used actual skulls on which to exhibit their artistic tendencies. Skulls were decorated with precious stones, etched with geometric designs, or replicated in intricate carvings. And we can't leave out the famous crystal skulls that have been attributed to the Maya, the Aztec, and even artistic aliens before finally being debunked as modern hoaxes. Apparently, our fascination with blinged-out skulls is a universal phenomenon.

Most of you will naturally associate the skull with its close cousin, the Jolly Roger. This emblematic flag sports a skull hovering above two generic bones (what the hell are they, anyway?) and is meant to instill fear in the hearts of those who cross paths with a pirate. And what fate awaits their victims? A little thievery, a bit of rape, and a likely stroll down the plank. But hey, everyone loves a pirate.

In reality, many a pirate ship sported a plain black flag, but the Jolly Roger has become the mainstay in pirate ship symbolism, which is understandable. It's a whole lot more interesting than your basic black and there's something extra creepy about the blank stare of a skull. Plus, it pays to have a catchy calling card when establishing your reputation as a first-class pillager.

The skull and crossbones have also served as warning for many a poison. Just seeing that symbol on a bottle evokes fear in the hearts of consumers. Be it rat poison, arsenic, or some backwoods, toxic moonshine, slapping a skull on the label is a surefire means of warning the thirsty.

The skull has also come to symbolize the badass. After chasing down pirates for a few hundred years, the military adopted the symbol, sporting it on flags, ships, and tattoos to signify their fighting prowess. The trend quickly caught on among the public, especially within the biking community, and today it's hard to find a biker who doesn't sport a skull somewhere on his person. From tattoos to tee shirts to leather jackets, bikers simply love the skull, which is ironic when you consider how many of them refuse to wear helmets. Perhaps it's time they design their own bony emblem—a skull with a hideous crack down the middle.

Skulls were also used to mark the entrance to ancient cemeteries. In the days of yore, when literacy was in short supply, the skull was a handy means of saying, "Enter at your own risk," for many a goblin was known to hang out in cemeteries. The skull served as a ghoulish reminder of the inherent dangers of the dead.

But the cultural fascination with our skeletons is not limited to our skulls. Bones have worked their way into our language, cropping up in sayings, slang, and nursery rhymes. Bone idioms (from the Latin, *idioma*, meaning a special phrase or expression) are widespread in English. And as someone who specializes in the human skeleton, these sayings give me great joy. So, I thought I'd share a few with you.

Got an argument to make? Then you have a "bone to pick." Still disagree? It becomes a "bone of contention." You can be "cut to the bone," "chilled to the bone," "feel it in your bones," or "work your fingers to the bone." You can possess a "funny bone," a "jealous bone," a "crazy bone," or be a "bag of bones." Someone can "break your bones," "throw you a bone," or (if you're lucky!) "jump your bones." And speaking of sex, we can't leave out that most common of usages, the ultimate, "boner."

Let's face it, our bones are some of the coolest parts of our bodies, so it's no wonder we have woven them into our culture. Whether they are warning us of danger or painting a verbal picture, make "no bones about it," skeletons rock!

THE ANTIQUITY OF AGGRESSION
April 10

At some point in our lives, each of us will have the urge to strike another human being. Ironically, for me, that moment occurred when I was attending a symposium on the evolution of ethics. The symposium was hosted by the weird and whacky folks in the philosophy department and featured an array of papers on the evolutionary basis of moral behavior. It was fascinating. That is, until a particularly long-winded academic began spewing jargon-laden oration like a linguistic volcano, some of which he admittedly concocted for the purpose of his argument. As he unsuccessfully defended his theory, I sat amidst his bewildered colleagues and all I could think about was punching him in the face.

Later, when I had cooled down (aided by a hefty dose of gin), I got to thinking about my irrational reaction, which got me thinking about aggression in general. Why are we aggressive? Wouldn't the world be a better place if we were all peace-loving, tree-hugging, tofu-munching hippies? Maybe. But then again, perhaps aggression played a critical role in humans becoming human. Let's explore.

Aggression is part and parcel of the range of emotions displayed by humans. Anger, happiness, sadness, and empathy aid us in navigating our social spheres and are critical components of human interaction. But aggression was fundamental to humans' eventual domination of the planet, for without a bit of aggression, we may never have achieved complex civilization.

All of nature competes to some degree. The phrase, "survival of the fittest," coined by Herbert Spencer but often mistakenly attributed to Darwin, was a simplistic way of saying that those who compete more successfully (thereby leaving behind more offspring) will most likely nudge out their less successful counterparts. And humans are no different. We, like other animals, have struggled to

270

survive throughout our evolutionary history. Fortunately, through a series of lucky anatomical and physiological twists, we evolved a bigger brain, which gave us an unprecedented edge over our competition.

When we first stumbled upon the nutritional benefits of meat, it was most likely as timid scavengers, fighting for the best scraps. As our technology progressed, those simple clubs used to fend off fellow carnivores developed into efficient weapons that not only protected us from said carnivores but allowed us to add them to the menu. And the more aggressive we were as hunters, the more meat there was to go around.

Our aggression would have naturally been directed toward each other. Competition for resources, be they hunting territories, water holes, or mates, would have compelled humans to compete. Let's face it, you're not going to get the girl by simply squatting outside your cave, hoping she wanders by. Aggressive males would have had more opportunities to mate (as they do today), thereby outbreeding their docile comrades. But aggression wasn't restricted to males. Aggressive females would have been more successful at protecting their young, attaining provisions, and going after those aggressive males.

As populations expanded and communities gained complexity, aggression would have enabled some to rise to power, others (the docile) to occupy the lower strata. As societies grew, so did their need for resources, and the best way to acquire resources is by conquering your neighbors. Once again, aggression wins. This pattern not only held for the ancients, but it still holds today. Human history is littered with the corpses of the conquered, and the powerful have never achieved power as shrinking violets. They do it through sheer force.

Aggression may even be hardwired, for we know it originates in the amygdala, that small clump of neurons located deep in the brain that also plays a role in fear and pleasure. Experiments have shown that our good friend, dopamine—that lovely biochemical that rewards us during sex—is also triggered during aggression. Like they say, it's a thin line between love and hate.

Scientists are still trying to tease out the complex relationship between the brain, its neurotransmitters, and the genes responsible for human aggression. But one thing is for sure: without aggression, you probably wouldn't be sitting here, reading this blog. The culture in which you live, the society in which you thrive, was built upon the shoulders of aggressors.

The meek shall inherit the earth?? I don't think so. . .

BEAT IT!
April 17

If you have ever had to hunt for a parking space on a college campus, you know what a hellish nightmare it can be. The other day, I scored a primo spot, tucked beneath the shade of a giant oak, just outside the College of Music. I cranked down my windows and was reviewing my lecture, when suddenly I noticed a gangly little dude with a snare drum setting up shop in front of the building. The drumming quickly commenced and within minutes, his incessant *Rat-a-Tat-Tat* attracted a second band member—of all things, a cymbals player. Together, they banged and clanged their way across my last nerve.

Despite their annoying performance, it got me thinking about drumming. What is it about the banging of drums that gets our blood pumping? Imagine music without drums (OK, classical music aside). Tough to do, because drums play a critical role in the way music stimulates our bodies and, more importantly, our brains. And what blissful stimulation it is.

Drums are one of our most ancient musical instruments. Think about it. It doesn't take much to produce a drum (or drummer, for that matter. I've seen chimps pound out a decent rhythm with nothing but a branch and a coconut). Just about any surface can be transformed into an instrument. Wood, metal, skins, or gourds all produce their own distinct sounds. And whether they are played with sticks or hands, drums form the backbone to music, setting the tempo, tone, and ambiance of a song.

But it's that rhythmic beat that our brains find irresistible. Neurologists have discovered that rhythmic beats actually cause our brain waves to match tempo. A fast, pulsing beat drives our brain waves to keep time. A slow, methodical rhythm lulls the brain, which is why drums are used to induce meditative and trancelike states.

It turns out this wave-altering mechanism may be just the trick for treating conditions such as attention deficit disorder (ADD). In fact, scientists used rhythmic sound and light stimulation on a group of young ADDers and found it to be just as effective as medication in improving concentration and elevating intelligence scores (although ADD and a drum set could make for a lethal combination, at least for the parents).

And speaking of drums and brains. . . It turns out kids who play drums may have a leg up when it comes to intellect. Little drummers were shown to have improved IQ scores following a series of lessons. And the same holds true for adult percussionists. Studies have found correlations between intelligence and rhythmic ability. Those with the best rhythm tend to score better on intelligence tests, for it turns out the parts of the brain used for rhythm are also employed for problem solving.

Rhythmic therapy may also improve cognitive function in the elderly and folks with brain injuries. The stimulating effects of rhythm actually increased blood flow to the brain, which improved cognitive scores among a group of senior subjects and, therefore, may have application for victims of stroke and head trauma.

But the positive effects of drums go far beyond the individual. They are an essential part of the human experience, a fundamental aspect of culture. Drums served as efficient forms of communication among many African cultures and were effective means of transmitting messages over long distances. And drum circles are an ancient tradition spanning the globe, drawing people together for ceremony, celebration, and socialization. In fact, you would be hard pressed to find a culture that doesn't include some form of drumming. And what a boring and monotone culture that would be.

Drums are part of our human heritage. They allow us to communicate in a universal language; one that lifts our spirits, moves our bodies, and even manipulates our brains. So, perhaps that annoying snare drum player was simply indulging in a force greater than himself. If so, then rock on, little dude!

HOW TO STOP A BULLET
April 24

Americans sure love their guns. Nothing says the "US of A" like the stars and stripes, Mom's apple pie, and an AR-15. The latest tally boasts around three hundred thousand guns in the United States—and that's just handguns. If you throw in their cousins, the rifles and shotguns, the numbers soar to the millions (110 and 86, respectively). Yes, Americans sure love their guns.

I must admit, I'm a gun owner. I keep a .357 at my bedside (to scare off unwanted midnight callers), and a .22 hidden in the kitchen (to ward off a baking ambush). When I moved away from home, the first thing my father handed me was a gun. He came from a long line of responsible gun owners and, to tell you the truth, as a female living alone, I don't feel safe unless there's a gun within easy reach. I admit, it's a sickness.

But gun ownership has taken on frightening dimensions in the U.S. The latest craze (Freudian slip) is personal body armor. Gun enthusiasts are no longer satisfied with owning assault rifles, now they want to sport Kevlar vests, to boot. Seriously?? I know life can be dangerous, and everyone has the right to protect himself, but if you feel the need to wear a vest, perhaps you should consider a new hobby, move to a better neighborhood, or seek counseling.

Today's body armor is a manufacturing marvel. These high-tech vests sport state of the art materials and lightweight construction and can be easily concealed beneath clothing. But this wasn't always the case. Body armor, like the weapons it protects against, has evolved through the ages. Let's take a quick tour.

The earliest forms of protection were animal hides. Over two thousand years ago, the Chinese prepared for battle by strapping on the skins of rhinos, which not only protected against clubs and arrows, but would have also doubled as top-notch rain gear. Pacific islanders wove coconut palm fibers into protective garbs since rhinos were in short supply and coconuts plentiful. The clever Greeks carried bronze shields into battle, while warriors in Central America wore quilted armor, which protected them from weapons, but unfortunately was no match against smallpox.

As metallurgy evolved, so did our means of protection. Chain mail, linked rings or wires made from a variety of metals, was developed around 400 BC in the present-day region of Ukraine. The trend quickly spread and before long, these metallic garments were seen throughout Europe, Asia, and parts of Africa. Scale armor was also in vogue. These overlapping plates were made from metal, leather, horn, or bone and were as effective as the rhino gear, minus the ticks and the stench. But the pinnacle of armor emerged around the 14th century when the invention of the crossbow necessitated a bit more protection. Thus, full body armor was born, and these well-protected combatants, decked out in their fancy tin cans, clanked their way to victory.

But everything changed with the introduction of gunpowder. It began in 9th century China, where clever alchemists mixed saltpeter, charcoal, and sulfur into an effective concoction to treat skin infections. The fact that it also exploded was a serendipitous sidenote (although not so much for the patient). Its healing properties aside, this magical mixture was quickly adopted by armies, who packaged it in bombs and mines and merrily blasted their way around the globe.

The first "hand cannons," as guns were then called, were used in 1364. A wick was set ablaze, which touched off the gunpowder, which finally launched the projectile—in those days, a small but lethal metal ball. These were cumbersome weapons, requiring reloading each time they were fired, a difficult task amidst the frenzy of combat. Regardless, they spread throughout Europe.

But as civilizations fought their way to power, weapons evolved to keep pace. Within a few hundred years, through the invention of flintlock ignitions, rifles, and, later, Samuel Colt's revolutionary revolver, guns were everywhere, especially in the New World. There they quickly subdued native populations (albeit assisted by some pretty lethal pathogens) and ushered in that most gun-worshiping period in American history, the Wild West.

That Wild West mentality remains engrained in the American psyche, for guns have become a symbol of freedom, independence, and (according to many Republicans), the epitome of American culture. And with the proliferation of guns comes the need for better protection, especially for law enforcement officers tasked with patrolling our gun-laden streets. Today's lightweight vests, which combine high-tech polymers, typically Kevlar, woven together into materials five times stronger than steel, provide a vital layer of protection against a criminal's bullet. And since cops should be entitled to a technological edge when it comes to fighting crime, bulletproof vests should be restricted to crime fighters.

We often speak of the "evolutionary arms race"—that process that fuels natural selection, ushering in novel adaptations as species struggle to survive. That race is run not only by our genes, but by the cultures that define us. And as weapons evolve, so too do our defenses. Let's just hope common sense can keep pace.

MODEL BEHAVIOR
May 1

This may sound preposterous, but long before I became an archaeologist and, before that, a firefighter, I dreamt of being a model. Like many teens, I devoured the fashion magazines—Vogue and Elle were my bibles. But for me, it wasn't so much the fashion, it was more the amazing photographs of those beautifully svelte women.

Their body proportions were astounding. They were tall, lithe, and, most importantly, incredibly thin. Everything a young girl aspired to be. So, as I prepared for high school, I dieted like a fiend, losing the subtle layer of baby fat I had been toting since childhood, transforming myself into a willowy wisp of an adolescent. I was ready.

Despite sprouting to a meager five feet seven inches, I was signed by a top local agency and finally got a taste of my dream profession. But after several shoots and shows, I soon discovered that life as a model fell short of my visions of grandeur. Don't get me wrong, there was nothing like the thrill of the catwalk and the money was ridiculous, considering what little effort went in to strutting around in designer wear. It's just that it lacked purpose. Fortunately, college led me to paramedic school, which led me straight into the fire service. The only problem was it required another bodily transformation.

The thin frame I acquired for modeling was ill-equipped for the rigors of firefighting, so I set to work, running, lifting, and pumping my way to a muscled physique. And it's a good thing I did. It turns out maintaining the body of a model, in all its emaciated splendor, is anything but glamourous.

Let's start with what a healthy body looks like. The average woman should be composed of about twenty-two percent body fat. That's because fat plays a fundamental role in the body's metabolism. It provides a backup energy source when carbohydrates are scarce; it absorbs vital nutrients, such as vitamins A, D, E, and K; and it helps maintain proper body temperature. So, you can imagine what happens to an individual who lacks the necessary fat stores. And the situation is only compounded by the extreme dieting most models undertake in order to maintain their fat-free physiques.

The most serious culprits are those strutting the catwalk. Let's face it, clothes look best on tall, thin frames. That's why the average size on the runway runs between 0 and 2, hardly your typical body. So, for a model to compete, she must maintain a frame that fits the bill.

Those skeletal silhouettes are achieved through extreme measures, from drugs such as amphetamines, to the use of colonics and juice diets, to the tried-and-true method of simple starvation. And the older the model, the harder it is to keep the weight off. That's one reason the industry preys on youth—prepubescents come in smaller, lighter packages.

This industry not only harms the models, but it also wreaks havoc on the psyches of girls everywhere. The result? An epidemic of eating disorders, driven by the marketing of unrealistic body types.

Eating disorders affect around seventy million people worldwide—twenty-four million of those in the U.S. alone (and this statistic is from 2002!). Ninety percent of women with eating disorders are between the ages of twelve and twenty-five and, in a ghoulish survey, half the women questioned said they would rather be hit by a truck than be fat. Come on!!

Anorexia is the third most common chronic illness among adolescents, which is frightening, considering that same anorexic is about twelve times more likely to die an early death. In fact, about twenty percent of people suffering from anorexia will die prematurely due to complications that accompany this syndrome; typically, cardiac abnormalities or suicide. And the body issues that drive this disorder are creeping into an ever-younger age group. According to a study in the *Journal of the American Dietetic Association*, eighty-one percent of ten-year-olds claimed they were afraid of becoming fat, and over half the nine- and ten-year-olds surveyed said they felt better about themselves when they were dieting. *Time* magazine reported that eighty percent of all children have been on a diet by the time they reach the fourth grade!

Body image is a complicated issue, driven by many factors, most notably the media. Fortunately, the fashion industry is taking note. The Council of Fashion Designers of America has developed guidelines to address the issue of underweight (and underage) models on the catwalk. So, perhaps there's hope.

I guess I was lucky. By trading couture for a bunker coat, I avoided the mania of modeling and was rewarded with a fulfilling career as a firefighter. And now that I'm a bioarchaeologist, my build is truly irrelevant. The skeletons I work on couldn't care less what size I am.

THE EYE OF THE BEHOLDER
May 8

I want you to take a moment and imagine a beautiful human. Since this blog is read around the world, I'm betting the range of faces that come to mind is staggering. That's because beauty can be an elusive concept, and what constitutes it varies from place to place. It really comes down to culture.

Last week's *Model Behavior* got me thinking about the concept of beauty and I found myself flashing back to my childhood. I grew up reading *National Geographic*, flipping through its glossy pages, entranced by the exotic people staring back at me from those beautiful photographs. It taught me early on that humans come in a wide range of colors, practice radically different customs, and do some amazing things with their bodies.

But my adolescent brain was confounded. Why would people poke giant holes in their earlobes, chisel scars across their flesh, or insert pins in the least likely places? (My first glance at a penis pin nearly knocked me out of my chair!) Fortunately, as I matured and accumulated a few degrees in anthropology, I came to appreciate the fact that humans manipulate their bodies for various reasons, sometimes religious or symbolic, but mainly in their quest for beauty.

Take those scars, for example. In the U.S., we go to great lengths to minimize, erase, or conceal our scars. Yet if we take a quick trip across the globe, scars take on a whole new meaning. Among the Karo of Ethiopia, men sport scars to represent their warfare prowess. Karo women, on the other hand, do it merely for esthetics. These intricate etchings are considered alluring and represent a woman's sensuality. And Ethiopians are not alone, for scarification is found in many parts of the world, from the dusty Australian Outback to the lush island of New Guinea. The custom has even found its way into America for, it seems, tattoos are now only for the faint of heart. For the more daring among us, the needle has been supplanted by the blade.

Other forms of superficial beautification include tattoos and piercings, widespread customs believed to enhance the appearance of their hosts. Among the Maori of New Zealand, facial tattoos, known as "moko," not only represent tribal affiliation (and scare the bejesus out of their foes), but emphasize a woman's desirability. Like tattoos, piercings come in all shapes and sizes and can be placed just about anywhere on the body. Extreme piercings, such as ear spools, have been worn for over five thousand years, from China to Africa to the Americas, as status symbols. The women of Borneo have taken it one step further. By adorning their ears with weights, they stretch their lobes to unimaginable lengths.

And they are hardly alone, for ear stretching has shown up in some surprising places, from Egypt's King Tutankhamen to Ötzi, the five-thousand-year-old frozen Alpine mummy. Even the statues of Easter Island sport elongated earlobes.

Earlobes aren't the only body part that is elongated for the sake of beauty. Don't get excited, boys, I'm referring to neck rings. Probably the best-known practitioners are among the Kayan of Northern Thailand. Known affectionately as the "giraffe" tribe, Kayan women strive for beauty one ring at a time. The process begins around the age of five and continues into adulthood. Length is achieved not by stretching the neck but by flattening the collarbones, making the neck appear longer than it actually is. Like many forms of beauty, it is but an illusion.

Evolutionary psychologists have been arguing about the basis of beauty for decades. The universalists concede that culture plays a role in the perception of beauty, but they argue there are certain underlying fundamentals, such as facial symmetry, a clear complexion, and large eyes. In the opposite corner stand the relativists, who believe beauty, like other aspects of humanity, is a culture-bound phenomenon. How else to explain the bizarre expressions seen around the globe?

To me, it's something deeper. Sure, it helps if both sides of your face match, and big eyes and clear skin are always a plus. But there are some who lack all of these attributes and yet are still astoundingly beautiful. I believe it comes down to what's inside.

THE CIRCUMCISION DECISION
May 15

Back when I was a new paramedic, I worked at a Level 1 trauma center. As an inexperienced medic, I was relegated to the most menial of tasks: drawing blood, monitoring vital signs, and patrolling bodily fluids. But my least favorite chore was catheterizing patients.

After a few trial runs, however, I became quite proficient. I could cath a patient in the blink of an eye. One evening, a frail, elderly man wandered into our ER, complaining of urinary discomfort. As the catheter specialist, I was summoned. I prepped my equipment, explained the procedure, and then discreetly exposed his genitalia. And as I took hold of his member, I paused for a moment of awe. For the first time, I found myself face to face with an uncircumcised penis.

I steadied my poker face as I recalculated my strategy. There was a lot more skin than I was used to, and it took me a few seconds of floundering before my catheter found its mark. And as I advanced the tubing (imagine stuffing a straw through a sausage), I found myself mesmerized by his unusual appendage.

Little did I know, the uncircumcised are hardly unique. Like most things, it all comes down to culture.

Male circumcision goes back thousands of years. Historians still debate its origins, but most agree it probably had its roots in rituals surrounding purification. Since many cultures view sexuality as sinful, removing the foreskin may have served to rein in a man's sexual proclivities.

The most ancient examples come from Egypt, where historical accounts dating to over five thousand years ago describe the ritual.

The procedure is also recorded in bas relief and found in evidence on mummified remains. Although Jews adopted the practice early on, the Romans were rather fond of their foreskin, and passed laws to protect their precious prepuces.

In many cultures, there is great ceremony surrounding circumcision. The Jewish Bris (technically called a "Bris Mila," meaning "covenant of circumcision") is symbolic of God's promise that the Jewish people will live on, thus, its focus on that imperative male organ. It is traditionally performed by a *mohel*, someone specially trained in wielding a knife. Once the baby has been snipped, the guests are free to gorge themselves on the Seudat Mitzvah (aka, religious feast).

But circumcision is hardly restricted to the Jews. It is found all over the globe in varying frequencies; about one-third of all males worldwide. In the U.S., circumcision first became a medicalized practice around 1870 and, as hospital births became the norm, it became part and parcel to the medicalization of birthing, as well as a symbol of status.

In America, the rate stands around sixty percent, with slight variations based on race and ethnicity. In the U.K., about half of all male Londoners are circumcised, and the same holds true for Canada. In the Land Down Under, sixty-nine percent of Aussie-born males are circumcised, yet in nearby New Zealand, it's only around forty percent. In Sub-Saharan Africa, the overall rate is around sixty-two percent, and many circumcisions are performed later in life. The same goes for the Philippines, where over half of those circumcised are put under the knife in their teens.

But circumcision is falling out of favor, at least here in America, even though it affords certain health benefits. According to the American Association of Pediatrics, a circumcised penis is less prone to STDs such as herpes, HPV, and syphilis; there's a markedly lower risk of acquiring HIV; and it reduces the incidence of urinary tract infections and certain rare forms of cancer. Most of this comes down to hygiene. The less skin there is, the easier it is to keep clean. So, if you sport foreskin, be sure to scrub up. And regardless of your foreskin status, be sure to always glove up!

The decision to circumcise is not only based on medicine, but it is also highly social. Like many aspects of culture—what you eat, what you wear, the traditions you follow—circumcision has much to do with the group to which you belong. If most men around you are circumcised, chances are you will be too.

As for which is more attractive, it really comes down to personal preference. And just as we discussed in last week's post, beauty is in the eye of the beholder.

A MEDITATION ON DECAPITATION
May 22

As a natural extension of last week's piece on circumcision, I've found myself thinking about decapitation. During my years as a medic, I never had the opportunity to see a decapitation, although I glimpsed gory photos taken by comrades in the field: horrific wrecks in which the car and driver were transformed into convertibles; the motorcyclist who inadvertently raced his cycle beneath the hidden guide wire. There are many dangerous ways to lose your head.

Thoughts of decapitation get the philosophical juices flowing. Does the person feel pain? Is he aware of his surroundings? I've often wondered what the victim experiences once the head has been separated from the body. Perhaps nothing. But for the sake of argument, let's imagine for a moment what it feels like to be beheaded.

First, let's tackle the pain factor. Although there are over three million pain receptors throughout the human body, there are none within the brain. Thus, the brain itself cannot feel pain, which is why surgeons are able to perform brain surgery on conscious patients. But that doesn't mean a decapitation isn't painful, since there are plenty of pain receptors in the neck (ask anyone who's ever suffered whiplash). So, the decapitation would certainly elicit a painful response. But would the person be aware of the pain?

That's where consciousness comes in. At its most basic, consciousness is defined as the state of being aware of one's surroundings. Let's not digress into the philosophical theories of consciousness, for philosophy is like a wormhole: who knows where we'll end up. Let's stick to physiology.

There are many conditions that cause unconsciousness: low blood sugar, psychological stress, and abnormal heart rhythms, among others. But if we are talking about decapitation, we're strictly concerned with blood loss, for if the head is detached from the body, blood flow is no longer an option.

Although the brain can survive for up to six minutes after the heart stops beating, consciousness is another story. Since the brain cannot store oxygen, rapid blood loss means unconsciousness occurs in seconds. That's why you feel lightheaded if you stand up too quickly. The brain picks up on that drop in blood pressure and, as a result, you get dizzy. So maybe decapitation causes such an immediate loss of blood that unconsciousness is instantaneous. Then again, maybe not.

I'm hardly the first to wonder about the state of mind of the decapitated. The scientific literature is dotted with anecdotal evidence of eyewitness accounts describing facial grimacing, blinking eyes, moving lips, or a wandering gaze. Whether these are conscious movements or simply remnant neuromuscular twitching, we'll never know. But let's review the scant evidence, just for argument's sake.

Back when the French were still enamored of the guillotine, there was a natural curiosity about the experience of the beheaded. It is said that many of those sentenced to death were asked to blink, if they were able, once the guillotine had performed its duty, and there are supposedly reports that some did just that, for up to thirty seconds.

The most famous case is that of the criminal, Languille, who was sentenced to the guillotine for murder. A Dr. Beaurieux observed Languille's facial expressions immediately following his beheading, which the doctor then recorded in *Archives d' Anthropologie Criminelle*. They included the blinking of eyes and the movement of lips, which lasted for several seconds. When Languille's face relaxed, Beaurieux yelled his name and the eyelids slowly rose. Languille focused his gaze on the good doctor before his eyes slid closed again. Beaurieux repeated the exercise and was rewarded with one final, purposeful stare before the eyes glazed over and Languille was declared dead.

Now that beheading has gone the way of the firing squad, we may never know if the decapitated are aware. Of the few beheadings that still take place—namely the barbaric displays by terrorists—we can't help but wonder about the victims.

Certainly, they feel the pain of the blade, for the methods employed are hardly humane. But once the head is severed, are the victims still aware? Do they experience a fleeting sense of the barbarian standing over them? God, I hope not.

The Body Blog

BALD AND BADASS
May 29

I grew up the daughter of a bald man. The funny thing is, I didn't realize he was bald until one of my elementary schoolmates pointed it out. To me, he was just "Dad." His hair was irrelevant.

My father started losing his hair when he was very young. I've seen pictures of him in his twenties and, even back then, the balding was well underway. But he never seemed to mind. In fact, I cannot recall him ever griping about his baldness. What remained of his hair he kept tightly clipped. He wasn't one for drastic measures. No ridiculous comb-over, no magic potions. As a Navy captain, he had more important things on his mind.

Baldness is a big issue among men, probably because about sixty percent of them will lose their hair. You would think, over time, they'd simply accept it and move on. But telling a man to disregard his baldness is like telling a woman to ignore the aging process. Impossible. We are culturally programmed to lose sleep over such issues. So, let's take a moment to explore the realm of androgenetic alopecia, aka, male pattern baldness.

Aside from certain medical conditions, if you are losing your hair, you probably have your genes and hormones to blame. And unfortunately, there's not a whole lot you can do about either one. You are what you are, and your genotype was predetermined before you ever shot from the womb.

As for the hormones, here's how they work. Male pattern baldness (MPB) occurs in men who have a predisposed sensitivity to the hormone, dihydrotestosterone (DHT). I'm sure you recognize the "-testosterone" base of that term. That's because DHT is a form of male sex hormone. In men with sensitivity, the DHT acts like a toxin on the hair follicles, starving them of nutrients, causing them

291

to shrink, and eventually shutting down the hair's growth phase. And the areas most affected? Those on the top and sides of the head, resulting in the characteristic "horseshoe" pattern of MPB.

So, what's a guy to do? There are probably as many home remedies as there are bald heads on the planet. Most of them involve some sort of herbal concoction you massage on your gourd. There's licorice root, aloe vera, onion juice, and fenugreek (whatever the hell that is). If you have a sweet tooth, you can use honey, yogurt, banana, or cinnamon powder. Or, if you are a manly man, you can choose castor oil, black pepper, camphor, or snake gourd. I'm pretty sure the results will be the same. . . But instead of reaching for a remedy, perhaps you should consider the razor. It turns out, a shaved head says a lot about the man underneath.

First, let's take a peek at a few of the cultural manifestations of shaved heads. There are many situations associated with head shaving—not all of them good. But I like to accentuate the positive, so we'll breeze past the contexts of prison internment, lice infestation, and Nazi punishment and, instead, focus on the finer aspects of baldness.

In many cultures, shaving the head is a rite of passage, especially when it comes to religion. Buddhist monks prepare for the priesthood by having their heads shaved, a symbolic commitment to the Holy Life. Hare Krishna do the same as a way of renouncing materialism, although they may leave a tiny tuft on the back to distinguish themselves from their Buddhist brothers.

Many branches of the military require shearing of recruits and even the ancient Romans sported bald heads, although they tended to pluck instead of shave. (Ouch!)

But the best news for bald men comes from a recent study out of the University of Pennsylvania. The folks at the Wharton School experimented on people's perception of the shaved head, and the overwhelming response was that men with shaved heads were perceived as stronger, taller, more confident, more masculine and, finally, more dominant than their hairy counterparts.

So, if you are losing your hair, try embracing your baldness. It certainly worked for my father. Regardless of his baldness (or perhaps because of it), he was a singular badass.

BODY DOUBLE
June 5

Imagine, for a moment, what it would be like to be attached to another human being. Think about it. . . every day, every night, every moment spent linked to another person. I've always been fascinated by conjoined twins. I remember seeing pictures of them when I was a child, marveling not only at the day-to-day logistics of such a setup, but that nature could actually produce something so spectacular. So, let's take a look at this most unique phenomenon.

Conjoined twins are rare. They occur about once in every two hundred thousand live births, and around half are born dead. Of those who are born alive, about a third survive for just one day. It's a small minority who live on.

Girls have a better chance of surviving than boys. Doctors are not sure why. Although the chance of twinning is higher among males, females are about three times more likely to survive than boys. Thus, females make up about seventy percent of all living conjoined twins. So how does it happen? How are conjoined twins made?

Conjoining occurs when a single fertilized egg fails to divide completely during the first week of conception. The process that would normally produce two identical twins is halted for some reason, and the partially separated egg continues on its developmental pathway. The result? Two bodies fused as one. And the fusion can occur anywhere on the body, resulting in an array of amalgamated individuals.

The most common are joined at the chest. "Thoracopagus" twins make up about forty percent of the conjoined and usually share a heart, which can make surgical separation tricky. Another common form is those connected from the waist to the breastbone. "Omphalopagus" twins are similar in design to thoracopagus and account for about a third of all twins. The number of shared organs can vary, depending on the degree of conjoining, but it's not unusual for them to share livers, digestive tracts, and genitourinary systems.

"Parapagus" twins share a body and sport two heads; a condition most intriguing when you think what it must be like to share a single body. And then there are the "craniopagus" twins, those joined at the head, which can range from superficial attachment (bone and tissue) to the sharing of a single brain. They make up a very small percentage of conjoined twins but arouse the most fascinating contemplation. What would it be like to share a brain?

Take Krista and Tatiana Hogan, two eight-year-old brunettes who are joined at the head. As craniopagus twins, they are unique. Only a fraction of this type survive yet not only are they remarkable as twins, but they are also remarkable for the manner in which their heads are attached.

The conjoining of their brains has produced an unusual condition: the girls share a thalamus. The thalamus is a lobed organ within the brain that processes the bulk of the sensory signals received by the brain. It plays a role in controlling the motor systems responsible for voluntary movement and coordination, but it is also an essential aspect of consciousness. And this is where things get truly interesting when we're dealing with craniopagus twins.

Doctors who have studied Krista and Tatiana believe their unusual neurological arrangement enables them to share sensory input through what the docs have termed a "thalamic bridge." And what this allows the girls to do is share sensations.

For instance, if one of the girls takes a sip, the other is compelled to swallow. If one is pricked in the finger, the other grimaces in pain. When they were infants, a pacifier in one mouth had a soothing effect for both. The list goes on and on.

Throughout history, conjoined twins have evoked curiosity, fear, and even scorn. The 16th century French surgeon, Ambroise Paré, believed conjoined twins were the result of God's anger and the Devil's influence. But his notions were cloaked in the ignorance of his time.

For me, conjoined twins reflect the remarkable range of human expression, the amazing variability of embryological development, and the beautiful complexity of nature.

I SEE DEAD PEOPLE
June 12

It happens as it always does, whenever I pass through that intersection. As I cross the lanes of traffic, I think about the dead girl.

I'm in Orlando for the weekend, visiting family and friends, returning to all the places I love in the city where I've spent so much of my life. I moved here when I was twelve, left when I was thirty-six, and spent thirteen of those twenty-four years as a firefighter-paramedic before moving on for a PhD. And I find when I come home, the city is haunted with the ghosts of my past.

As I navigate the city's streets, scenes from my life as a medic flash before me in vivid detail. For instance, that intersection I mentioned. I was a new medic, working grueling hours on the ambulance, when a young pedestrian tried to cross the many lanes of traffic, only to be taken down by a semi. Her body was defleshed from the waist down, a condition known as "degloving," and I can still remember the heat radiating off the pavement, the stench of the truck's smoldering brakes, and the roar of nearby traffic as we wrapped her shredded limbs.

There's the schoolyard where my partner and I performed CPR beneath the gawking gaze of schoolchildren. The sidewalk where I tried in vain to staunch the flow from a suicidal gunshot wound to the head. The strip mall, where the young man set himself on fire. And the trauma center to which I delivered countless patients, victims of the city's unrelenting violence. My brain is a virtual card catalog of tragedy.

I know *why* this happens. These memories are part of my past, forever etched into my psyche. The bigger question is *how* it happens. How does the brain pull forth memories buried deep within the subconscious? It turns out neuropsychologists have been hard at work studying the processes involved in memory retrieval.

It all starts with retrieval cues: clues or prompts that trigger the brain to recall information. But it's the type of retrieval cue that determines just how the brain pulls forth long-buried events.

Recall involves the straightforward retrieval of information, such as answering a simple question. There's little work involved, and the information simply pops up when prompted. *Recollection* requires a bit more effort. Your brain reconstructs the memory by pulling together bits of information, such as clues or partial memories, reassembling them into a greater whole.

Recognition retrieval occurs when your brain latches on to something familiar, like selecting your favorite dish from a menu. And finally, *relearning* is just as it sounds—retrieving information you have learned on some previous occasion, which often results in stronger memories that are easier to recall. But where do these memories reside?

The brain stores memories in one of two ways. Short-term memories are processed in the prefrontal lobe, that clump of brain located just behind your forehead. The short-terms are translated into long-term memories in the hippocampus, a small horseshoe-shaped structure within the limbic system that rounds up memories from various sensory regions in the brain and binds them into a single memory episode. Over time, the neuronal connections associated with that memory become fixed and can be replayed at will.

The hippocampus also helps solidify the connections that form our memories, and each memory serves as an index for our recorded thoughts and sensations. Through functional MRI, scientists have been able to observe the brain as it reconstructs memories. And as each memory is recalled, different regions of the brain light up as various sensations and thoughts are replayed.

So, when I return to Orlando, memories of my patients reemerge, triggered by the sights, sounds, and smells of the city. I recall their faces, their injuries, and their pain, and once again I experience the intense emotions they evoke whenever they resurface in my mind. These memories are relics from a previous life, carved into my subconscious, and forever part of who I am.

> *"Memories are bullets. Some whiz by and only spook you.*
> *Others tear you open and leave you in pieces."*
> *- Richard Kadrey, Kill the Dead*

MY FATHER'S DAUGHTER
June 19

Here's a thought: if the sperm that created you had come in second in the race to the egg, you would be an entirely different person. Think about it. Among the million or so sperm vying for that egg, the one that contained the recipe for you won, and if any other tadpole had gained entry, you would not be who you are. Quite a gamble, procreation.

As Father's Day approaches, we naturally think of our dads. Which got me thinking about the process of conception and the traits a father passes on to his children. Man, was I lucky.

First and foremost, there's the brain. I was fortunate to have a very intelligent father, and he exploited his intelligence to achieve great things. He was born and raised in Mississippi, the son of a prominent architect. But when he was still a boy, his father left to start a new family, leaving him and his mother behind. So, he dropped out of school and went to work.

Realizing the grim future in store for a Mississippi kid with a ninth-grade education, he enlisted in the Navy when he was just sixteen. And once he left the south, he never looked back.

That young, uneducated sailor went on to travel every corner of the globe, complete a master's in theology from Northwestern University, and achieve the rank of Captain. All of it through sheer force of will.

He is the reason I pushed on for a PhD. In fact, he's the reason for much of what I've done in my life, for he represented the pinnacle of success, the example of what hard work and hard-headed determination can achieve.

But conception is like life; we must take the bad with the good. Aside from determination, he also passed on some less-than-desirable traits. For instance, that same hard-headedness, which can border on stubborn; a deep and abiding love of gin; and above all else, his temper.

My father's temper could go from zero to sixty in a matter of seconds; a trait which in my case has fortunately mellowed with time. Although my father was a chaplain and a man of God, he could swear like any seaworthy sailor. Politics, the economy, or criminal activity would set him off and he would launch into a tirade, cursing all of civilization, damning human weakness.

But he was also one of the funniest individuals I've ever known. He loved to laugh, and my fondest memories are of sipping martinis and listening, enthralled, as his stories unwound. Like when his ship was torpedoed, plunging him and his crew into the dark depths of the Pacific. And how his commander had silenced him with, "*Shut Up Sailor, We'll Get to You!*"

Dad was also a philosopher. His worldview was an intriguing blend of religion, philosophy, and the hard life lessons of his youth. He had several famous sayings, such as, "*If it's worth doing, it's worth overdoing,*" a creed he employed whenever he ordered Chinese takeout. Or his other motto, "*Why do today what you can put off until tomorrow?*" which is ironic, considering all he achieved in his lifetime.

He spent forty-two years in the Navy and survived two wars—a feat no man can endure without being forever altered. But his cynicism was balanced by a love of laughter, a warmth of spirit, and a clever mind that never failed to see the humorous side of life.

My father is gone now. A slow-growing tumor bloomed deep within his frontal lobe, dimming, and eventually snuffing out, that most vibrant of personalities. I still see him in the mirror. I share his eyes, the shape of his face, and his strong, white teeth. But his most important traits reside within me, for he graced me with a curious mind, steely determination, and a will that has sustained me through every crisis in life; one of the hardest being his death.

EIGHT-LEGGED ENVY
June 26

This morning, on my way to my car, I walked through a gargantuan spiderweb. A diligent arachnid had been hard at work, industriously spinning his beautiful web, only to have some bumbling human destroy it in one fell swoop.

Of course, I had my hands full since these incidents never occur when one is unencumbered. I tried swiping the web from my face, only to realize its owner had conveniently hopped aboard my person. I suddenly became aware of a chunky spider the size of a malt ball taking a leisurely stroll down my arm.

Despite the burden of my computer, a coffee mug, and my purse, I managed to flail my limbs with enough vigor to dislodge him. He gracefully sailed down his web, landing gently at my feet and then scampering off into the undergrowth. After giving myself a thorough rubdown to ensure I wasn't toting a giant egg sack on my back, I gathered my belongings and went on my way.

I'm not particularly afraid of spiders. I hold them in the same regard I hold snakes: cautious respect and deep admiration for the way they ambulate. Image what humans would be like with eight legs instead of two? It would probably render automobiles obsolete.

My eight-legged encounter got me thinking about our own mode of locomotion. In the animal kingdom, walking on two legs (bipedalism) is pretty unique. Only two other bipeds readily come to mind—penguins and kangaroos, both of which have devised their own strategies for getting around. Penguins have sacrificed efficient walking for swimming, and kangaroos took to hopping, which sure beats walking across the Australian bush. So, why did humans evolve such an unusual gait? Perhaps we should first ask, "when?"

Ancient fossils are hard to come by. The older they are, the less chance they have of being preserved intact. But there are clues to bipedalism among the fragmented remains of our earliest ancestors, and some of the best evidence has nothing to do with legs.

The hole in the base of the skull where our spine enters is called the foramen magnum. And it's the position of this hole that provides a clue to upright walking. When it's oriented at the base of the skull, it shows that the creature stood upright. If the hole is located toward the back of the skull, it indicates a quadruped (think about your dog or cat).

And it turns out our bipedal gait evolved much earlier than once believed. It was once thought that walking on two legs evolved in concert with our large brains. But what we find in the fossil record is that bipedalism was in place millions of years before our big brains arrived on scene. Even the seven-million-year-old *Sahelanthropus tchadensis*, unearthed by a group of French paleoanthropologists in 2001, is believed to have been bipedal, based on his foramen magnum. Although scientists are still quibbling.

But the bigger question is *why*? Why did humans switch from four to two legs? Theories of bipedalism go all the way back to Darwin, who believed the freeing of our arms allowed us to concentrate on the production of tools and weapons. This makes sense until you take into account that stone tools don't show up until many millions of years after we started scooting around on two legs.

Others believe climate change had a hand in it. Perhaps humans took to walking as their forests were reduced and food became harder to come by, prompting males and females to partner up for provisioning. Males could gather food (in their arms, of course), and provide for their female and offspring, which would cement their bond and benefit both parties.

Or perhaps the reduction in forests required our ancestors to traverse longer distances. Walking upright or, better yet, running, has been shown to be more energy efficient than the knuckle-walking of our primate cousins, and there's a whole new line of research examining the role running may have played in the evolution of humans.

Whatever the reason, we humans wouldn't be human without our unusual gait. Sure, spiders have it made, what with their eight legs and their ability to walk on water. But it's hard to imagine how we humans could have accomplished all we accomplished if we were still ambling about on all fours. Stone tools, pottery, weapons, and art would have been quite a challenge without free hands, as would carrying, whether it be food, firewood, or children.

So, you can keep your eight legs, Mr. Spider, and I'll stick with my two. Your arachnid abilities may grace you with unusual gifts, but it only takes one of my two feet to squash you like a pancake.

HOOKING THROUGH HISTORY
July 3

No exploration of the human body would be complete without a brief glimpse at that most ancient of professions, prostitution. How can we possibly explore the body without contemplating the sale of said body? So, let's go back in time and trace the evolution of this infamous trade.

As long as man has wandered the planet, I'm sure some form of prostitution has been in place. It comes down to simple supply and demand. I can just imagine a consensual agreement involving sex in exchange for a juicy mastodon shank or some handy work around the cave, for we all know a way to a man's heart is not necessarily through his stomach.

Prostitution has many euphemisms, more so for women than for men. Male prostitutes are typically gigolos or hustlers. Females, on the other hand, sport a rash of labels, most of which are hardly flattering. Hooker, streetwalker, whore, and skank are among the most common. In the days of yore, prostitutes were known as strumpets, trollops, harlots, or courtesans. But regardless of gender, history is riddled with accounts proving tricks have been turned for thousands of years.

In the ancient Near East, the Sumerians (conveniently) wove prostitution into their religion. Religious prostitution in Babylon required women to venture forth to the sanctuary of Militta at least once in their lives to have a conjugal confrontation with a foreigner; all in the name of hospitality, of course.

Prostitution among the Greeks was common among women and young boys. In fact, the Greek word for prostitute is *porne*, which is derived from the word meaning, "to sell," laying the groundwork for what thousands of years in the future would become a thriving industry.

The first Greek brothel was opened in the 6th century BC, with earnings going toward building a temple dedicated to Qedesh, the patron goddess of commerce. They even had categorical names for the various types of prostitutes, depending on where they worked, be it on the streets, in houses, or near bridges (don't ask me). As for male prostitutes, they were quite popular among the Greeks, and the profession was usually taken up by adolescent boys, slave and free alike.

The Romans believed in farming their prostitutes and would round up abandoned children and raise them for future sale. Slaves were also captured in battle or purchased for the sole purpose of prostitution, and sex for sale was even used as a form of legal punishment for women.

By the Middle Ages, the Roman Catholic Church was on a rampage to tamp down the trade in Europe, although it received blowback from those who believed the service helped prevent rape, sodomy, and (god forbid!) masturbation. Most brothels were left to their vices, as long as they resided on the outskirts of the village. That is, until cities caught on to the popularity of red-light districts, where clients could window-shop for whatever caught their fancy.

But things changed in the 1490s following the return of Columbus' voyages to the New World, for hidden aboard his cargo lurked a deadly stowaway: syphilis. Syphilis became widespread throughout Europe, with prostitutes serving as popular hosts for the bacterium. This only added fuel to the fire of reformationists bent on tearing down this illicit trade. And even though folks were experimenting with various types of condoms, from catgut to sheep bladders, those rudimentary rubbers were no match for the "pox."

By the 19th century, France, followed by the U.K., passed laws to ensure regular medical examinations for their prostitutes. The Contagious Diseases Act mandated regular pelvic exams for their "pros"—not only on home turf, but also in their colonies abroad.

Of course, around this time in America, prostitutes were as common as cattle among the dusty plains of the Wild West (and treated equally as well). Whoring was one of the few professions available to women of the period and, as America spread westward, so did prostitution. Wherever a new town popped up, so did a brothel that would set to servicing the menfolk, lickety-split.

But by the early 1900s, the buzzkill organization known as the Woman's Christian Temperance Union marched in and quashed not only brothels, but alcohol, to boot. And by 1917, even New Orleans' famous Storyville district—sixteen blocks of unfettered frolicking named after the councilman who established it—was closed, despite public outcry from the locals. One had to travel all the way to Alaska to buy a legal poke in those days.

Today, Nevada is the sole host to legalized prostitution in the U.S. About thirty brothels support around five hundred prostitutes who work as independent contractors, most without the need for a pimp. As for the rest of the world, it's a patchwork of legal and illegal selling. In a survey of one hundred countries, prostitution was illegal in thirty-nine, somewhat legal in twelve, and legal in forty-nine others.

And fierce debates abound about whether it should be a legalized profession, with advocates claiming legalization protects its practitioners, and women's rights groups claiming it is inherently abusive.

Despite its legal status, prostitution is part of human culture and, as history goes to show, wherever there's a demand for sex, there will always be someone peddling it.

FIRE DOWN BELOW
July 10

Last week, we took a brief glimpse at the long and convoluted history of prostitution, so I thought it only natural to follow up with an infectious postscript.

For as long as humans have been exchanging bodily fluids, pathogens have been part of the mix. And when it comes to bumping genitals, there are a whole slew of contagions getting in on the action. Because a comprehensive overview is beyond the scope of this blog (not to mention my short attention span), we'll stick to the highlights while we explore the dark and daunting world of STDs.

First, let's clarify the terminology. You may have noticed that the term STD has lately been supplanted by STI. What differentiates a sexually transmitted *disease* from a sexually transmitted *infection* is the presence of symptoms. However, since many STDs never cause symptoms, it's really splitting hairs. So, for the sake of today's post, we'll stick to the tried-and-true acronym, STD.

Sexually transmitted diseases most likely evolved along with humans, and historians have been chronicling their presence all the way back to the Bible. The Old Testament refers to "the running issue," referencing the "clothing needing washing as did the man himself," most likely referring to gonorrhea, which causes that telltale discharge from the penis. And it wasn't until the Middle Ages (around AD 1200) that the disease was finally linked to sex. But gonorrhea is only one of many STDs plaguing humans, for the list of potential pathogens is long and varied.

STDs come in three basic varieties: bacterial (gonorrhea, syphilis, and chlamydia, to name a few); viral (Hepatitis B and C, Herpes, HPV, and HIV); and parasitic (such as trichomonas, a pesky protozoan that thrives within urethras and vaginas). And it's the type of pathogen that determines the treatment.

Since the advent of antibiotics, the bacterial bugs can usually be wiped out with a simple prescription, as can trichomonas. Unfortunately, the viral pests are not so simple. Once a person is infected, herpes and HIV are here to stay, and one can only mitigate the symptoms. Hepatitis, however, forms a mixed bag. With Hepatitis B, most people can be cured, although a minority will become carriers for life. Hep C holds a more dismal future, as a majority will suffer long-term infection with chronic liver disease on the horizon.

But try to imagine what these maladies must have been like before the advent of modern medicine. So, to keep things in perspective, let's peruse some of the ancient treatments that were once believed to cure the "fire down below."

The ancient Greeks were some of the first to record the treatment of venereal disease. In fact, the term "herpes" originates from the Greek, meaning "to creep or crawl." And how did they attack the creepy crawlers? By burning off the lesions using hot irons. Despite their torturous treatments, they get kudos for instituting public policies aimed at reducing the spread of herpes, although their "no public kissing" rules probably did little to curb the virus.

By 1746, London's Lock Hospital was the first to establish public treatment programs for the infected. And the 18th and 19th centuries saw the use of mercury, arsenic, and sulfur as the primary remedies, although these dangerous regimes caused serious side effects, even death. Despite the danger, arsenic, in the form of Salvarsan, was used to treat syphilis well into the 20ᵗʰ century.

And as scary as these diseases can be, what scares people even more is the social stigma attached to them. However, for those of you harboring an STD, take heart. You are hardly alone in your affliction. Here are a few statistics to bring it all home.

According to the CDC, there are over three hundred million new cases of STDs in the world each year. The human papillomavirus (HPV) is now the fastest growing STD and nearly all sexually active folks will contract it at some point in life.

About one in five Americans has genital herpes, yet about ninety percent of them don't know they have it. And health officials warn that by 2025, up to forty percent of men and almost half of all women could be infected with this permanent virus.

And of course, HIV is still rampant, still spreading, and still deadly. As the sun sets in South Africa, another fifteen hundred new infections will have taken place today. Yes, I said fifteen hundred *per day*. And that's the conservative end of the statistic.

Let's face it, STDs are scary, and the emotional toll they incur can be as burdensome as their symptoms. But pathogens, like us, are thriving members of the biome and will forever be a part of life on our planet. So, stay informed, stay healthy and, for god's sake, use a condom.

PLAYING DEFENSE
July 17

Last week, my body came under attack. In an ironic twist following last week's post on contagious pathogens, I picked up a nasty bug that for the past seven days has wreaked havoc on my immune system. Fortunately, whatever I caught was confined to my northern regions—primarily my throat and chest—rendering me febrile, voiceless, and with a bone-rattling cough that could give any tubercular a run for his money.

I'm happy to report that I am now on the mend, but it got me thinking about the immune system and the vital role it plays in keeping us safe. Naturally, I thought I would elucidate its magical machinations, but I found myself resorting to boring military metaphors traditionally employed for such discussions. The trusty lymphocytes that serve as armed forces, always on high alert and ready to mobilize should a foreign invader appear on the horizon. Pathogens, those dangerous usurpers who are just waiting for the opportunity to bust through our protected borders. Blah, blah, blah.

So, instead of the usual immunity song and dance, I thought we'd explore the more perceptible means of defense, for our bodies have evolved numerous nifty ways to rid themselves of unwanted guests.

First and foremost is that largest of organs, the skin, which accounts for around sixteen percent of our body weight. Skin serves as a protective barrier against our pathogen-infested world and it does this not only through its layered arrangement, but also by producing specialized peptides that annihilate microbes and sound the alert when danger approaches.

But there are two problems when it comes to skin's defenses. First, skin tears. And once it is torn—whether through an injury, an insect bite, or on purpose, through surgery—it allows entry to all sorts of dangerous organisms, from bacteria, to viruses, to parasites.

The second problem concerns topography: although our skin is one large organ, it varies from surface to surface, and some of our most vulnerable surfaces are those that house our mucous membranes. For example, the respiratory tract. The moist, gooey surfaces of our respiratory system provide the perfect portals for pathogens. Each time we put a hand to our mouth, pick our nose, or simply take a breath, we can usher in a suite of infectious organisms that would love to plant their flag.

Fortunately, our respiratory tracts have devised a few clever ways of ridding themselves of pesky pests, which explains why we cough, sneeze, dribble, and blow. Our lungs also sport a thin layer of microbe-fighting proteins, which defend against any bugs that manage to weather the snotty storm.

But pathogens are crafty. Some, like influenza, actually attach themselves to our bronchial membranes to prevent their quick expulsion. Others, such as measles and whooping cough, render our cilia inoperable. Those small, hair-like projections are designed to usher pathogens up, up, and away from our lungs, and when they are knocked off-line, bugs can simply run rampant.

The respiratory tract is but one of many portals for pathogens. Our stomachs are prime targets for many food- and waterborne bugs, which cause a wide range of misery, illness, and death. Luckily, our stomachs make for fairly acidic accommodations, with an average pH of about 2 (which you science nerds will recognize as pretty darn acidic). And just like our respiratory tracts, our gastrointestinal plumbing has devised a couple of rapid evacuation methods, namely vomiting and diarrhea.

And speaking of acidic body parts, let's not forget the vagina. This acidy little tube sports a pH of around 4, which is ideal for warding off bacterial and fungal invaders—not to mention sperm, which explains their desperate swimming. Not so, our urethras, which is why urinary tract infections are so common. Especially in women, for not only do our urethras lack defenses, but they are positioned dangerously close to the anus, which as we all know is a virtual playground for pathogens.

And speaking of that other southerly portal. . . The anus, like the urethra, is also ill-equipped to ward off infection. And what makes it even more dangerous is that, unlike the vagina, the anus lacks any natural lubrication. So, if you are going to use it for recreational purposes, do yourself a favor and lube up. It will prevent tissue tears, which are great access points for infection. And don't forget the condom!

So, the next time you find yourself sneezing, coughing, vomiting, or worse, take a moment to appreciate the fundamental necessity of such functions and know that, as miserable as these symptoms are, they serve a vital role in the fight against pathogens.

THE FEAR FACTOR
July 24

If you had to list the things that scare you the most, what would your list include? Human fears run the gamut, from the insignificant (spiders and heights) to those that haunt us in the wee hours of the night (loss of a loved one, inevitable death). Fear, like other emotions, is a visceral aspect of humanity. But it goes even deeper than that, for fear transcends the boundaries of humanness. It's part of our evolutionary heritage.

Want to scare a chimp? Place a plastic snake next to an unsuspecting primate (humans included) and you'll probably witness pure, unadulterated fear. That's because the fear of snakes appears to be hardwired into many primate brains; a deep-seated phobia that may have evolved to keep us safe. Since many snakes possess the ability to kill, it seems only logical that animals that avoid a close encounter might have an evolutionary edge over the less cautious.

But where does fear reside? And what actually happens when we are scared? Like any emotion, it all begins in the brain. Many parts of the brain are activated during the fear response. And the majority of them are located deep within, a testament to their ancient origins. Yes, our fancy cortex also plays a role in fear, but the rest of the hardware we share with other animals, since critters lacking fear would stand little chance of surviving in our dangerous world.

Here's a quick glimpse at the brainy bits responsible for processing fear. Our sensory cortexes interpret what we see, hear, smell, and feel. The information is whisked to the thalamus, which decides where to shuttle the data, and the hippocampus then places the data in context. The amygdala decodes the data and determines if a threat exists. And if the threat is real, the hypothalamus activates the "fight-or-flight" response, which kicks the body into high gear to respond to the situation.

Of course, these reactions happen with lightning speed and, in many cases, the body simply responds as if threatened, even if the threat turns out to be benign. It's better to gear up than to sit back and contemplate. A momentary hesitation could spell death.

The hypothalamus activates two separate systems when it launches the fight-or-flight response. The sympathetic nervous system activates stress hormones, adrenaline, and noradrenaline, which are dumped into the bloodstream. As they circulate, they increase heart rate and blood pressure, which explains the thumping in your chest that accompanies a scary jolt. At the same time, the pituitary gland gets involved by secreting a hormonal cascade that primes the body for action. Pupils dilate to improve visual acuity, blood vessels in the skin constrict to shunt blood to the major muscles, and muscles tense for action (which explains the goose bumps). While the essential functions are enhanced, nonessentials, such as digestion and immunity, are sidelined. That way, the body can focus on the immediate threat and conserve energy in the process.

But if fear evolved to improve survival, why is it so many of us love a good scare? I admit, I'm an adrenaline junkie, much of which I blame on the years I spent as a firefighter. Once you've rushed headlong into a burning building, daily life can seem a bit monotonous, which probably explains my love of rollercoasters, skydiving, and scary movies.

But the reason many of us love a good scare is because the fight-or-flight response involves many neurotransmitters; namely endorphins, dopamine, and serotonin, that are also responsible for a rush of pleasure (think orgasms). That is why a momentary scare is followed by a blissful blast of relief. Once our brain realizes the fear isn't real, our body can simply enjoy the rush, which is why screaming is often trailed by nervous giggles.

But humans can do something no other animals can: they can conjure fear. Our sophisticated brains enable us to do some amazing things. But they also come at a price, for although we are gifted with imagination, much of our imagining can evoke fear.

Fear of the future, fear of loss, fear for the ones we love. . . there are a million ways we torment ourselves by conjuring fear. But it is worth noting that, however much we languish in fear, it has little effect on outcome.

So, keep your fear in check and save it for life's true emergencies. The next one could be right around the corner.

SOUL SEARCHING
July 31

In the two years since I have been writing this blog, we have explored just about every aspect of our anatomy, from the beautiful intricacies of its form and function to the bizarre ways we modify and even mutilate our bodies. So, as this writing endeavor draws to a close and I focus on other projects, I thought I would end by discussing an aspect of the body that has eluded scientists and philosophers for centuries: the quest for the human soul.

Throughout history, the soul has been part of our search for understanding how the human body works. Ancient terms to describe the soul—from Latin's *anima* to the Greek *psyche*—usually refer to the vital forces within the body, be they motion, movement, or breath. And since the presence of the soul was believed to separate the living from the dead, it seemed only logical that it should reside somewhere in the body. All we had to do was find it.

Some of the earliest references to the soul go back over five thousand years to the Egyptians, who believed the soul was composed of five parts, the most important of which resided within the heart. The heart was believed to be the animating life force, the source of our feelings, thoughts, and will. In fact, the weight of the heart at death determined the soul's destiny. If the heart were considered too heavy, it would be consumed by a demon, subsequently ending one's bid for the afterlife. This cardiocentric view of the soul persisted throughout much of history.

Fast forward a few thousand years to the Greek poet Homer, who claimed there were two types of souls. The first, which resided somewhere in the chest, controlled our emotions, everything from joy to reason to rage. The second type of soul was tied to a person's individual identity and appeared only in dreams. It had no specific location within the body but served as the animating life force. Homer believed it was this aspect of the soul that fled the body at the time of death.

The foundations of Western philosophy, forged by the likes of Socrates, Aristotle, and Plato, also contemplated the riddles of the soul. Plato considered the soul to be of celestial origin, the immortal essence of a person that was divided into three parts. The rational aspect, which controlled reason, was of primary importance and thus located within the brain. The spirited aspect, responsible for courage, resided in the chest, and the appetitive portion, which governed love (of food, drink, and "loving delights") was located in the abdomen. The goal of life was to achieve a balance within the soul, especially regarding spirits and appetites; a human struggle that continues to this day.

Plato's student, Aristotle, stoked his own ideas about the soul. He agreed the soul formed the essence of an individual but, unlike Plato, Aristotle believed the soul could not be separated from the body. So much for its immortality. . . He too divided the soul into three parts but, in his view, the soul controlled bodily functions and was therefore defined as such: the vegetative function (nourishment and reproduction); the sensitive function (sensation and movement); and the intellectual function (cognition and deliberation). Aristotle also believed that all animals possessed a soul, although the intellectual functions were confined to humans. And like the Egyptians of long ago, Aristotle believed the heart served as keeper.

Early Christians took a broader view. The soul not only gave form to the body but could be found in every aspect of our anatomy. It was believed the soul entered the body only after the fetus was fully formed. "Delayed ensoulment" coincided with the "quickening," thus once the mother felt the baby move, the soul was considered to have arrived.

Around the seventh century AD, as the Dark Ages blanketed humanity, the belief in delayed ensoulment persisted. The Roman Catholic Church decreed abortion acceptable as long as it was carried out before the soul arrived, and this was upheld well into the 19th century.

With the blossoming of the Renaissance in the 1300s, Leonardo da Vinci incorporated the search for the soul in his anatomical studies, declaring the middle ventricle of the brain as the most logical spot. René Descartes took up the banner a few hundred years later, agreeing with Leo on the general location of the brain, but claiming the pineal gland was a more likely location.

As scientists learned more about the inner workings of the brain, belief in a craniocentric soul persisted, well into the 20th century. It seemed only natural that the seat of consciousness should also house the soul. But as science advanced and our understanding of the human body crystalized, the soul as animating life force slowly fell away. The mystical realms of life could now be understood in terms of biochemistry, neurology, and genetics, and issues of the soul were gradually relinquished to the theologians.

If you ask me if we possess a soul, I'd have to say I don't know. The scientist in me embraces the tangible explanations for what constitutes a living body, and I'm far more comfortable discussing cellular respiration than arguing the validity of delayed ensoulment.

But that in no way diminishes my fascination with life or the wonder I feel when I contemplate the intricacies of our anatomy. Regardless of our beliefs, we can all agree the human body is a truly astounding machine, one that not only sustains us but enables us to experience our world.

As for the ghost in the machine. . . I'll leave that to the theologians.

ACKNOWLEDGEMENTS

This book is based on a blog that began over a cup of coffee between friends. I want to thank Michael Boonstra, for explaining what exactly a blog is and for planting the seeds for what would become a two-year-long writing endeavor. I also want to thank my gifted editor, Chris Gallaway, for not only gracing this manuscript with her talent and skill, but for making each of my books that much better through her involvement. And finally, my wonderful family, for loving and encouraging me along the way. Thank you all so much.

Thank you for reading *The Body Blog: Explorations in Science and Culture*. I hope you enjoyed it.

Since reviews are an author's lifeblood, I would sincerely appreciate it if you took a moment to **leave a review** at your place of purchase.

To stay up to date on my books/publications, please go to:

- https://rachelwentzbooks.com/
- Like my Facebook page: https://www.facebook.com/RachelWentzBooks
- Follow me on Twitter: https://twitter.com/RKWentzBooks

For those interested in archaeology, please check out my other books on Amazon:

Life and Death at Windover: Excavations of 7,000-Year-Old Pond Cemetery

Chasing Bones: An Archaeologist's Pursuit of Skeletons

Searching Sand and Surf: The Origins of Archaeology in Florida

Or for fiction readers, my award-winning novel, *The Mass of Men*, which won a Silver Medal in the Florida Authors and Publishers Association annual President's Award in Genre Fiction and is now an IndieReader "Best Of" book. You can check out the review here: https://indiereader.com/book_review/the-mass-of-men/

The first chapter is provided below.

The Mass of Men

She stood in line, the lone female among thirty-four cadets, an interloper, branded by a lack of testosterone. The cadets braced at attention, feet planted shoulder-width apart, fists knotted at the base of their spines as the instructors made their way across the polished wood floor, their footsteps echoing off the high walls of the gymnasium.

The heat in the gym was palpable, although the instructors seemed immune, their uniforms crisp, impeccable. The cadets sweated freely, their shorts and tee shirts, emblazoned with the academy logo, clung to their torsos as the vast space was infused with the odor of anxious flesh.

Samantha stood erect, lengthening her spine in a futile attempt to gain an extra inch, flexing the muscles of her back to appear bigger and stronger than her hundred and forty pounds. She stared straight ahead as the instructors worked their way down the line, examining the cadets with slow, critical eyes. They moved deliberately, shoulders back, stomachs in, remembering with nostalgia what it was like to be young, naive, and scared shitless.

Sam winced as they looked her over. She wasn't femininely beautiful, she sported the fresh-scrubbed looks that belonged in a J. Crew catalog, more innate than obvious. She had inherited her mother's lovely legs, her father's strong white teeth and hardheaded determination. Her hazel eyes were handed down from some distant German relative, as was her blond hair, which she had cropped boy short. Sam could feel the instructors taking in her shape, their eyes sliding down the length of her legs. For the first time since her struggles through puberty and braces, Sam was embarrassed by her looks.

As two instructors circulated in front, a third stood off to the side, taking the students in from a distance, sizing them up from afar. The instructors were seasoned officers, well-versed in the world of firefighting. They relished their role of breaking new recruits.

The students ranged in age from teenage boys fresh out of high school to men who had ventured down various career paths, some as professional athletes, others in the military. The older cadets were less intimidated. It showed in the way they carried themselves – erect stance, eyes forward. The eyes of the younger men danced in their heads. Somehow, they had all ended up together, grappling with their insecurities, waiting for the instructors to complete their excruciating inspection.

The taller of the officers reached for his clipboard, which had been placed with precision on a nearby chair. A whistle lay neatly beside it, its cord bundled in a tight coil. The officer maintained perfect posture, even while bending forward. He snapped back to attention as his eyes slid down the list of names, his booming voice sending a nervous ripple through the line of cadets.

"When you hear your name, respond with a 'check.' Ables! Adams! Bradford! Chancellor…"

Each name elicited a resounding "Check!" When he called out the name Peterson, a sarcastic "Yes, sir!" emanated from a big lug near the middle of the line. The officer paused as the second instructor swooped in.

"Did you not hear the instructions, or are we now admittin' retards?" shouted the burly officer in a southern slur, nose-to-nose with the reckless cadet.

"Y-yes," the lug stammered, "I mean, no."

"You better tack a 'sir' onto that or you'll be kissin' my fuckin' boots!" the officer barked, as the student shriveled with indignity.

"Kissing the boots" was code for pushups. Twenty pushups were the academy's standard reprimand, the paramilitary equivalent of a ruler to the back of the hand. For egregious errors, the whole group would perform.

Peterson dropped his eyes before muttering, "Yes, sir."

The officer backed away as roll call continued.

The instructor made it to the end of the alphabet. "Smith!"

"Check," Sam responded in a deep voice, a vain attempt to disguise her femininity.

Three more names were called, and roll was complete. The officer swung his clipboard behind his back, locking it between powerful hands as he braced before the cadets.

"Welcome to Class #64 of the Central Florida Fire Academy. I'm Officer Michaels," he said in a stilted tone meant to convey the seriousness of the situation. Sam noted his shaved head on which she could just make out a faint ring of auburn hair. He wore a brand-new baseball cap with the academy's logo; scant protection against the Florida sun that had intensified the rash of freckles covering his face and arms. A bushy handlebar mustache of darker auburn covered his mouth and stood in relief against a powerful jaw. His body was athletic, every inch of it tense, ready to spring.

"This is Officer Tanner," Michaels continued, with a quick jab of his clipboard.

Less polished but just as powerful, Tanner sported a dark shock of hair contrasting eyes of brilliant blue. He dangled his thumbs from his belt loops as he slowly raked the line. His thick frame disguised a dense layer of muscle underneath and a worn toothpick appeared permanently lodged in the corner of his mouth. He cocked his head as he fingered the pick, rolling it against the faint callous on his lower lip as he eyed the cadets.

Tanner stepped forward, removing the toothpick to mark the gravity of the moment. He threw back his shoulders as his deep drawl ricocheted off the gymnasium walls. "You're ours for the next three months. During that time, we'll lose a third of you, especially the heavy kid and the girl," he bellowed, his blue eyes skipping over Sam as if she had already been dismissed. He tossed a nod to Michaels and then plunked the toothpick back into his mouth, chewing for a moment before maneuvering it back in place.

Sam froze, embarrassed to be singled out so early in the process, furious at being deemed unworthy. She felt the old familial anger rise, an attribute passed from father to daughter, triggered whenever anyone dare challenge her resolve. She kept her eyes forward, all the while imagining ramming the toothpick through Tanner's fleshy jowl, yelling, "*Fuck you, sir!*" as she snapped a rigid salute.

She swallowed her anger, instead focusing guilt-ridden relief on the overweight kid who had decided, against all better judgment, to try to make it as a firefighter. Sam knew he might deflect attention from her, that his lack of physical strength might draw the wrath of the instructors, and that his inadequacies might make her transition easier. She stole a glance down the line and could just make out his trembling belly protruding among the row of washboard abs.

The third instructor, the one standing off to the side, glided slowly into her field of vision. Sam's spine stiffened as he took his place before the group. He was well built, lean, and appeared to be in his mid-forties. He had sandy hair cropped short and piercing grey eyes that sailed up and down the line before locking onto hers for a brief second. A matching mustache hid most of his mouth – a faint slit cut into the sharp contours of his face. Sam noticed he was handsome, but the coldness of his eyes overrode his good looks. She tried to imagine him smiling. He stood before them, shoulders squared, hands locked behind his back, examining his new class of cadets.

"This is Commander Daniels," Michaels announced reverently before ducking his chin and backing away.

"I welcome each of you," Daniels said in an even, quiet tone as his grey eyes slid up and down the line. "I hope you've come prepared to work. The officers are here to assist you through your training, so if you have a problem, let them know. If there's something you can't do, let them know. If you decide you've made a mistake by being here, let them know. Don't waste our time and we won't waste yours."

With a glance he turned the floor back to Michaels and resumed his position on the periphery of the room. Sam continued to brace as Michaels rifled the pages on his clipboard and laid out the day's schedule.

~

Commander Daniels watched her from the side of the gym. He had been watching her ever since she had taken her place in line. He tried not to, but the sense of familiarity was consuming. It settled in a portion of his brain where calm and pleasure once resided.

Perhaps it was the blond hair, the intensity of her eyes. Or perhaps it was the determination in her face that caused his mind to turn back on itself. Daniels forced his eyes forward, keeping the girl in his periphery as he tried to focus.

As Michaels wrapped up the overview, Daniels stole another glance and was startled by Tanner's barking order, "Take five and get back in line!" The words jolted the commander back to the present as his reverie disintegrated. With one last glimpse, he turned and marched slowly from the gymnasium.

~

After the break, the students filed back into the gym for the fitness assessment, which would gauge their strength and endurance. Here they would prove they possessed the stamina to enter a burning building wearing the sixty pounds of gear that would protect them from extreme temperatures and falling debris.

Once the cadets were lined up in formation, they were shuffled outside for a two-mile run. Three more instructors emerged from the gym, their faces hidden beneath ball caps and sunglasses, black tee shirts stretched taut across their muscled chests. They fell in behind the cadets and unleashed their fury.

"At this pace we'll be here all goddamn day!" one of them yelled as the officers pursued the cadets like a pack of jackals, hell-bent on culling the herd.

"Pick up your fuckin' feet!" another screamed as the officers infiltrated the group, darting among the students, disorienting them with their swiftness.

Sam stayed with the front of the pack, having trained for the last six months as she completed her paramedic certification. The officers exploited her strength, using her as an anvil to crush the weaker cadets.

"The girl's kicking your asses!" an officer yelled at the cadets bringing up the rear. *"Pick it up, you bunch of pussies!"* he screamed, dashing among the stragglers, his verbal assault aimed directly at their ears. A startled cadet stumbled and went sprawling to the pavement. The officer screeched to a halt, his stopwatch gripped in a sweaty fist.

"Since you can't keep up with the group, you fuckin' turtle, you can kiss my goddamn boots!" he barked as the cadet rolled into position and began hammering out pushups.

Sam knew the academy would be grueling; she wasn't about to show up unprepared. The heavy kid hadn't planned as well. He was huffing and puffing his way to the finish line as Tanner kicked gravel from the sideline, his clipboard flapping at his side like a broken wing as he shouted, *"Get your big ass movin'!"*

With the run completed, the students staggered back to the gym for the rest of the assessment. They jockeyed into formation as they struggled to catch their breath.

"Give us as many pushups as you can. The max score is 50. Start on my count," Michaels announced as the cadets broke from formation and spread out within the gym.

Sam dropped into position, noting the faint smell of socks permeating the wood floors, listening to the grunts of the men nearby. The cadets braced in plank position as the instructors stood by, relishing each painful minute before finally blowing their whistles. By then, the weaker individuals were tiring, their backs slowly drooping, bellies kissing the floor. Sam focused on a far spot on the wall and pumped out fifty with little trouble.

Sit-ups were next. They had ninety seconds to perform. Sam hammered out eighty as one of the bigger cadets watched, straining to match her pace. But he had a massive upper body to contend with, Sam, the advantage of a lighter frame. The stronger cadets did well. The weaker ones wobbled from side to side, grimacing. Sam watched as the chubby kid squeaked out twenty. The instructors took note of her efforts. Tanner even managed a brief smile when he grumbled, "Nice job."

Following sit-ups, the students fanned out, arms and legs spread-eagle as a team of assessors tackled them with a vicious metal clasp. They measured the students' body fat by pinching the flesh along the backs of their arms, their waists, and the insides of their thighs. Sam noted the unsteady hand of the assessor as he placed the clasp against her leg. She stifled a grin as he scowled down at his clipboard, scribbling furiously.

Sam turned her attention to the heavy kid who stood nearby, still panting from the run, sweat rolling down his pink face as the calipers strained to accommodate his layers.

"Better set 'em to maximum!" the big lug from roll call snarled as those nearby snickered.

Michaels was on him in a flash. He braced before the cadet, his green eyes blazing. "I see I'm going to have to make you my special project, Peterson," he whispered irately, their faces close. "You can start by kissing my boots."

Peterson's shoulders slumped and he exhaled loudly, then dropped into position in front of the officer. Michaels pivoted; the toes of his boots positioned directly beneath Peterson's face. As the cadet pumped out twenty pushups, his forehead tapping the officer's boots with each downward stroke, Michaels took the opportunity to address the students.

"The academy is a paramilitary institution. You are expected to adhere to our code of conduct at all times," he proclaimed from a Superman stance as the cadet bobbed at his feet.

Peterson finished his pushups and sprang upright with a grunt. He brushed his hands off on his shorts and grudgingly resumed his place among the students. Michaels eyed the group and moved off.

When the pinching was done, each cadet thoroughly welted, they were made to squeeze a metal-handled contraption to assess their grip strength. The more macho in the bunch used the opportunity to show off, bearing down on the handle as they maxed out the gauge. The cadets watched closely as Sam squeezed out a modest score. Peterson shot her a belittling snort.

The heavy kid stepped up, glancing nervously at the group before setting his face and straining against the handle. As he gripped with all he had, bending forward to force his energy into that miserable contraption, a loud, whining fart escaped, sending the rest of the group into nervous fits of laughter. The instructors quieted them with quick shouts. Sam turned away, smiling despite herself.

With the assessment completed, the cadets were hustled to the showers. The men moved in a sweaty pack, swearing softly so as not to be overheard by the instructors. Sam broke from the group, relieved to be out from under the instructors' scrutiny, relishing the silence and isolation of the women's locker room, which was immaculate from lack of use. She stripped out of her damp clothes, jumped under a cold stream, washed with lightning speed, and ran a towel over her body before wrestling her way into a tight sports bra. She appraised her flattened chest with a quick sweep of her hands. She then donned a crisp, new uniform – black lace-up boots, matching fatigues, a blue academy tee shirt, and a heather-blue dress shirt, starched to perfection. The uniform felt foreign, her reflection in the mirror that of a stranger. Her short hair only added to her boyish appearance.

The cadets emerged as newly minted clones. Their dress shirts sported the academy patch on one sleeve, an EMT or paramedic patch, if certified, on the other. Their chests boasted silver name tags, last name only, as if their lowly rank afforded only partial identification. Their badges would be distributed upon graduation. Thus, began their first day at the academy.

~

The students sat bolt upright at their desks, speaking only when spoken to, writing frantically as Michaels outlined their curriculum. He braced in front of the class, broad shoulders flexed, his hands gripping the podium as if it might take flight. His bicep bulged as he flipped to the next page.

Tanner stood off to the side, leaning his large frame against the wall, eyeballing the cadets as he chewed on his pick. Commander Daniels moved silently throughout the room, gazing down on the students' notes, pulling papers from their desktops at random to gauge attention spans, notetaking skills. He stopped for a moment beside Sam's desk. Sam could smell the leather from his belt intermingled with an earthy soap smell. She stole a sideways glance and could make out the blond hairs that stood out on his forearm, which was tanned and corded. His hands were locked behind his back and Sam wondered if the man ever relaxed. She glanced up just as he turned away. Daniels sailed to the back of the room and disappeared.

Back in his office, the commander rifled through paperwork before leaning back in his chair, hitching a boot up on the open drawer, and gazing out his window. It seemed like a solid bunch. Most would pass. A few wouldn't make it. He had developed a keen eye for those who would succeed, those who would fail. He was rarely wrong. The evidence was on their desks.

Daniels could gauge a student's potential by the way they took notes. He recognized the frantic scribbling of those determined to stay on top. Then there were the others, those who scrawled pictures and geometric designs as they watched the minute hand perform its slow creep.

He was curious how the girl would do. He had seen the intensity in her face, but would that translate? She wasn't the first female to come through, though they certainly weren't common. Daniels ran a strict academy. His reputation for breaking recruits was well deserved. If a student wanted an easier path, they could always travel to the next county over for a watered-down version of his curriculum. Few women graced his door.

Her notes impressed him. They were neat and clearly written, each of her sections labeled and underlined; an attention to detail Daniels appreciated. He had noticed her hands, the right of which continued to work despite his presence, the left splayed out on her notebook, stabilizing it as she wrote. Her fingernails were trimmed short, no polish. Not the painted daggers of the last female, who ranted when told they would have to go. This one was prepared to work and, most importantly, blend in.

~

"You have thirty minutes for lunch," Tanner announced as he erased the whiteboard and swung his bulk to face the students. "Be back by thirteen hundred or your ass is mine." He grinned before yelling, "Break!"

The students scrambled from their desks and headed to the cafeteria in a noisy herd. Lunch finally afforded them a chance to talk. Sam slid into the seat next to Johnny Simms, a paramedic she knew from the months spent completing her clinicals. Johnny smiled before turning his attention to a pile of wilted greens.

"So, when did you decide to give up the ambulance for firefighting?" Sam asked as she plowed into her lunch, grimacing at the blandness of the vegetables. She dropped her fork in disgust and bit off a hunk of dry biscuit, chasing it with a gulp of water.

"When I realized it was the only way to build a pension," Johnny replied with a wry grin.

Johnny's easy demeanor camouflaged a laser focus. He had a methodical approach to patient care and Sam considered him one of the most talented medics in the county. She glanced at the long scar that ran along the left side of his face. As she did, two cadets took seats on either side of the table. The taller of the two spoke up as he vigorously shook a carton of milk, his voice modulating with the jerking of his arm.

"I'm Lance, this is Trey," he said, indicating with a jut of his chin the leaner, sandy haired individual next to Johnny.

Sam noticed the dichotomy of their appearance: Lance was dark haired, lanky, with giant hands and feet, and a bushy mustache that clashed with his baby face. He cracked open the carton and took a deep swig, his Adam's apple bobbing with each swallow. Trey was lightly built, subdued and serious. He gave a stiff nod as he meticulously sliced his turkey. Where Lance's movements were awkward and erratic, Trey's were restrained, systematic. They seemed to play off each other, like an old married couple. Sam watched as Lance eyed Johnny's scar.

"That's quite a scar you got there," Lance mumbled in between bites as he continued to stare at Johnny's face, oblivious to his discomfort. "How'd you get that? Bar fight?" He grinned over a mouthful.

"Car wreck," Johnny muttered before dropping his eyes to his tray. Sam knew he would say no more. Johnny guarded the story of his wreck. All Sam managed to get out of him during her tenure on the ambulance was that it happened when he was a teen and he had gone into cardiac arrest. That made him a miracle, in Sam's eyes. Trauma arrests rarely came back.

"Where did you guys come from?" Johnny asked as a means of diverting attention from his face.

"Michigan," Lance replied, taking another deep chug from his milk. He swiped a quick finger over his mustache. "We were sick of freezing our asses off, so we applied and relocated. Just arrived two days ago. Trey's piece-a-shit Honda barely made it." He eyed Trey and paused to gulp a few more bites. Trey took up the story.

"We had a bit of trouble getting to class this morning," he added gravely, bracing his fork and knife against his plate as he recounted how his car had sputtered to a halt three blocks from the academy.

Lance took over, leaning into the conversation, working his arms for effect. His knife and fork hung in midair. "So, I'm standing in the intersection, trying to shove the car to the side of the road," he began, shifting his eyes conspiratorially. "Trey's screaming at me to push, cars are *whizzing* by us, no one will stop to help, and some guy leans out the window and yells, 'Hey assholes, go back to Michigan!'

Here Lance paused, wide-eyed. "Is that any way to treat a fella?" he whined before tucking back into his lunch.

"Don't sweat it," Johnny replied, grinning. "This isn't the best area to be looking for help." Johnny would know. He had worked the west side for years and was familiar with the brutality of the territory.

As they returned to their meals, a young cadet quietly took up the seat across from Sam. He smiled shyly as he flipped open his napkin. "I'm Tyler Williams, but I go by Ty," he murmured, cutting his eyes to his tray.

"I'm Sam," she replied, noting that his shyness was enhanced by the beauty of his face, the perfection of his bones. Ty was graced with exquisite coloring; the soft brown of his hair, the deep tint of his skin, and dark hazel eyes, which Sam could barely appreciate because of his dodging glance. After some coaxing, Ty told her about growing up in Ocala, and how it was assumed he would be taking over the family's sprawling farm.

"Why did you leave?" Sam asked after Ty fell silent.

"My father," he replied, eyes on his tray.

Sam noted the strain that played across his features and was struck with the urge to touch his face.

Sam had the alphabetical good fortune to be positioned between Johnny and Ty when they stood in formation. The heavy kid hadn't been as lucky. His first name, "Tate," had quickly been transformed by the officers to "Tater" on account of his spud-like physique, and his last name of Patterson put him elbow-to-elbow with the smartass, Peterson. He was doubly cursed. He sat in isolation, hunched over his tray.

Like Lance and Trey, Shane and Bo had also signed up together. They were young, both sons of Orlando firefighters, and both attempting the academy for the second time. The seriousness of their situation sapped any excitement they may have felt. The pressure, most likely doled out by their fathers, was plastered across their young faces.

Aaron Peterson sidled up to Sam's table with an exaggerated swagger, his belt loaded down with an array of tools and a knife big enough to take down a good-sized hog. His large head sported a fleshy face, and his narrow eyes hovered above a cocky grin. Sam watched as he plunked down his tray and cut his beady gaze to Ty. He settled into his lunch, continuing to analyze Ty, while scooping large forkfuls of mashed potatoes into his face, which didn't keep him from blurting out his thoughts on the academy.

"This is going to be a bitch!" he exclaimed as he worked a piece of bread around his plate, smacking his lips with gusto. "Had a cousin who went through a year ago, said Daniels is one cold motherfucker. Pluck you out just for lookin' at him wrong. And Michaels ain't much better," he continued in between gulps to no one in particular. "Got an ass so tight he can whistle Dixie."

When no one responded, Aaron took it as an insult and scanned the table for a victim. His gaze landed back on Ty.

Ty eyed him before returning to his tray. Aaron glanced from Ty to Sam. "Well, it looks like we got two pretty girls in class," Aaron announced, grinning and leaning back in his seat. "I only noticed the one this mornin'." He swiped at his mouth with a meaty hand.

Ty ignored him and stabbed at a hunk of broccoli. Sam glared down the table at Aaron as if examining a large turd, and then slowly swung her eyes to Ty. Her gesture only encouraged the lug.

"Hey, Pete," Aaron shouted at the lanky kid across the table, apparently a comrade from the sticks. "Which one's prettier? Hard for me to decide."

Pete forced a grin and kept eating, eyes on his plate. He was used to Peterson's blathering, having attended the same EMT course. He knew to smile and nod; the only way to avoid falling victim to Aaron's cruelty.

"Hey! Pretty-boy!" Aaron called down to Ty. "You sure you're in the right place? I think the fuckin' home-ec class is down the hall!" He laughed loudly despite a mouthful of peas.

Sam glanced over at Ty and saw the brown skin of his neck flush. She could see the anger rise in his face, but he maintained control and kept his eyes down. Although Ty was not tall, he had the build of someone used to the physical work of a farm. But he seemed disinclined to fight. Sam admired his restraint. Her mind was churning out ways to verbally pummel Peterson, but she knew a response in Ty's defense would only encourage Aaron's antagonism. Instead, she leaned in and whispered across the table, "Jealous asshole," bringing a warm smile to Ty's lips.

Aaron grunted his disapproval at their communal lack of humor and returned to his potatoes.

~

The "cigar sessions" had become a ritual. At the end of the first day, Michaels and Tanner congregated in Daniels' office to discuss the new group of cadets and go over class logistics. They gathered in the soft leather chairs, an arrangement Daniels implemented when he took over as commander of the academy, so that he could converse with his team without the barrier of a desk.

The cigars were a secret the officers shared. Prior to lighting up, Daniels would issue a quick nod and Tanner would spring from his seat and quietly close the door, since tobacco was forbidden at the academy. It was their singular guilty pleasure. They earned it.

The only person privy to their ritual was Miss Davis, Daniels' Administrative Assistant. Nothing got by Maggie. She knew everything that went on at the academy, no matter how covert the operation. To harass the officers, she always waited until she smelled the telltale smoke emanating from beneath the office door before poking her head inside. As soon as she appeared, the men dropped their hands in an attempt to conceal their contraband.

"I'm leaving, Commander, if there's nothing else you need," she said sweetly, forcing a straight face and batting crinkled, knowing eyes. Her plump frame was hidden behind the door and her round face seemed to float in midair.

"We're fine, Miss Davis," Daniels replied gruffly with a guilty wave of his empty hand. In the nine years Maggie had worked for him, he had never referred to her as anything other than Miss Davis. Intimacy, even in the slightest degree, did not come naturally to the commander.

Once the door was securely closed, Daniels relaxed in his chair and eyed his officers. "So, what do you think?" he began, drawing on his cigar and leveling an icy stare. The sessions always started with the same five words.

Michaels sat bolt upright, frozen at attention. Even in the comfort of the office, Michaels' posture remained impeccable. He flipped through his clipboard, which he kept perpetually at his side, attached by some invisible umbilicus. "A strong bunch," he replied as he scanned his notes. "A few weak ones, but they should be gone in no time. One smartass, but overall a stable group. I think they'll do well." As he spoke, he held a protective hand over his clipboard to guard it from falling ashes. He eyed Tanner for input.

Tanner was leaning back, puffing leisurely on his fat cigar, his toothpick temporarily dislodged and sticking out from his breast pocket. He relished the meetings in the commander's office. They were a reminder of how far he'd come since his days as a green cadet. He took one more savory puff before commenting.

"Pretty sharp group," he began, eyeing the burning end of his cigar. "Peterson's gonna be a handful, but nothin' we can't tamp down with a little pressure. Good lookin' chick," he added with a wink. It'd been a while since they'd had a female at the academy.

Then Tanner remembered.

Daniels forbade the instructors to comment on female cadets. Tanner was reminded by the razor look Daniels shot him as he reached for the ashtray. To cover the blunder, Tanner continued in a formal tone, parroting Michaels.

"I think they'll do well," he said, his twang barely audible as he stiffened in his seat. He looked to Michaels for a reprieve, but Daniels pushed through so as not to dampen the air.

"We've got some young ones in the group," Daniels continued, flicking his ashes. "Shane and Bo are back, so keep an eye on them and see they don't get in trouble." Daniels knew from experience that the difficult males, especially those hell-bent on proving themselves, tended to prey on the youngest. It always happened when men converged, as if the training weren't enough to prove their prowess. The young ones were often crushed in the process.

Daniels went over the schedule, assigning adjunct instructors for the days spent on the training ground. Safety regulations dictated one instructor per six students; training with live fire required an even higher ratio. Lining up the adequate number of adjuncts was a constant in Daniels' daily routine. He relied on Michaels to stay on top of it.

With their cigars smoked down to stubs and the schedule set, the meeting wound down and Michaels and Tanner rose to leave. Tanner popped his toothpick back into his mouth and grabbed the ashtray on his way out. It would be returned by morning, emptied and clean. It was just something he did, a form of tribute for being included.

With the office to himself, Daniels returned to his desk and scanned the list of cadets. His eyes slid down the names, settling near the end. *Smith, Samantha,* he read, thinking back to her appearance in lineup, berating himself for remembering. He rose from his chair, tossed the list onto his desk, ran a hand through his bristled hair, flipped the lights off, and left.

www.ingramcontent.com/pod-product-compliance
Lightning Source LLC
Chambersburg PA
CBHW020853180526
45163CB00007B/2491